T0230435

Mitigating Environmental Impact
of Petroleum Lubricants

Ignatio Madanhire · Charles Mbohwa

Mitigating Environmental Impact of Petroleum Lubricants

 Springer

Ignatio Madanhire
Department of Engineering Management
University of Johannesburg
Johannesburg
South Africa

Charles Mbohwa
Faculty of Engineering and the Built
Environment
University of Johannesburg
Johannesburg
South Africa

ISBN 978-3-319-81021-8   ISBN 978-3-319-31358-0   (eBook)
DOI 10.1007/978-3-319-31358-0

Printed on acid-free paper

This Springer imprint is published by Springer Nature
The registered company is Springer International Publishing AG Switzerland

We sincerely dedicate this book to our families and friends. Special mention goes to colleagues at both the University of Johannesburg and the University of Zimbabwe for their support and inspiration. Also to teams we worked with in the oil industry in Zimbabwe over the years.

Preface

This work reviews effective environmental impact mitigation for petroleum-based lubricants to reduce their negative persistence during usage and upon end-of-life disposal. The book explores the basic tribology of lubricants as well as initiatives that may enhance the environmental and economic effectiveness of lubricating oils from the composition design perspective. Reference is made to mineral base oil processing, blending, application and disposal of petroleum lubricants, and the book presents and extends current best practices that minimize or eliminate adverse environmental impact throughout the product's life cycle. The book also presents some in-depth insight into base oil/additive substitution, use of biolubricants in total loss application which are biodegradable, consideration of synthetic lubricants to extend drainage interval, use of quality bases in Group III and Group IV to achieve fuel economy and reduce emissions, rerefining of used oils, as well as recommending environmentally friendly disposal of used lubricating oils. Some effort was made to equip readers with technical understanding of lubricating oils' chemical and physical properties in terms of their potential hazardous nature to humans, acquatic species, water bodies and soil properties, where mitigatory initiatives were equally presented from base oil selection, additive development especially for total loss use. The book ends with a review of solid lubricants in severe space operations as the way forward to minimze environmental impact. Issues highlited are of benefit in terms of achieving both environmental legal compliance and eco-labelling business competitiveness—all the while preserving the environment for sustainability. It is in this regard that the book is therefore of interest to both manufacturers and consumers in the lubricants industry.

Acknowledgements

We wish to thank our colleagues at the University of Johannesburg in the School of Engineering Management for sponsoring conference travels for presenting the papers on related topics which later on inspired the authors to come up with this book. We are also grateful to Cuthbert Chidamba and Tineyi Mhundwa for the diligent and focused reviewing the material during the manuscript generation. Their interrogation that was supported by wide experience on the subject matter was just intense to nourish the progress of material compilation and rearranging.

Thank you all.

Ignatio Madanhire
Charles Mbohwa

Contents

1   **Introduction**.. 1
    1.1   Background 1
    1.2   Mineral Base Oil Processing 2
    1.3   Lubricants Blending.................................... 4
    1.4   Lubricants Types and Applications 5
    1.5   Impact on Environment 8
    1.6   Lubricants Handling.................................... 10
    1.7   Disposal and Harmful Ingredients 11
    1.8   Biodegradability and Toxicity 12
    1.9   Lubricant Life Cycle 13
    1.10   Conclusion 14
    References... 15

2   **Lubricant Additive Impacts on Human Health**
    and the Environment....................................... 17
    2.1   Introduction 17
    2.2   Environmental Effects of Used Oil 18
    2.3   Environmental Impacts................................. 19
    2.4   Effect of Used Oil on Soil 21
    2.5   Future Development of Eco-friendly Lubricants 23
    2.6   Chemistry of Lubricant Additives and Their Toxicology
          to Humans 23
    2.7   Lubricant Additives and Their Hazards to the Environment 27
    2.8   Ultimate Fate of Lubricant Additives..................... 30
    2.9   Biodegradation of Additives............................ 32
    2.10   Bio-concentration of Additives. 33
    2.11   The Future of Additive Technology 33
    2.12   Conclusion 34
    References... 34

3   The Environment and Lubricant Related Emissions.   35
    3.1   Introduction   35
    3.2   Extended Drain Interval   38
    3.3   Fuel Economy Aspect.   38
    3.4   Emissions Reduction Aspect   40
    3.5   Contribution of Lubricant Properties to Diesel
        Exhaust Emissions   42
    3.6   Lubricant Additives on Particulate Emissions   44
    3.7   Conclusion   44
    References   45

4   Green Lubricant Design and Practice Concept   47
    4.1   Introduction   47
    4.2   Vegetable Bio-lubricants   48
    4.3   Environmental Pollution Control   51
    4.4   Lubricants for High Temperature Diesel Engines   52
    4.5   Synthetic Lubricants and Long Drainage Intervals   52
    4.6   Additives to Match High Temperatures   53
    4.7   Lube Deposit Formation.   53
    4.8   Hot Metal Surface Effects   54
    4.9   Environmentally Considerate Lubricants (ECL)   55
    4.10   Recycling and Reclamation of Lubricants   56
    4.11   Extended Condition-Based Drainage Interval   56
    4.12   Leakage Management.   56
    4.13   Future of Green Lubricants   57
    4.14   Conclusion   57
    References   58

5   Synthetic Lubricants and the Environment   59
    5.1   Introduction   59
    5.2   Synthetic Versus Mineral Lubricants   60
    5.3   Synthetic Base Oil Classification   61
    5.4   Demand for Thermal-Oxidative Oils for High-Temperature
        Diesel Lubricants   62
    5.5   Lubricants Based on Synthesized Fluid   63
    5.6   Hydrocarbons Build-up from Mineral Oil-Based
        Lubricants.   63
    5.7   Health Impact on Humans   64
        5.7.1   Toxicity of Unused Lubricating Oils   64
        5.7.2   Toxicity of Used Lubricating Oils   65
        5.7.3   Effect of Extended Drainage Interval
                on Used Oil.   65
    5.8   Environmental Impacts.   66
    5.9   Advantages of Synthetic Lubricants.   67
        5.9.1   Fuel Economy   67
        5.9.2   Extended Drainage/Reduced Oil Disposal.   67

	5.9.3	Particulate Emissions Reduction	68
	5.9.4	High Temperature Stability	68
	5.9.5	Bio-degradability of Synthetic Lubricating Oils	69
5.10	Impact of Recycling Used Oil		70
5.11	Synthetic Lubricants and the Future.		70
5.12	Conclusion		71
	References		72

6   Eco-friendly Base Oils. **73**
6.1	Introduction	73
6.2	Bio-based Base Oil.	74
6.3	Lubricant Base Stocks	75
6.4	Eco-labeling of Lubricants.	75
6.5	Features of Good Bio-lubricants	76
6.6	Base Stocks from Vegetable Oils	76
6.7	Bio Lubricants Market	78
6.8	Make-up of Vegetable Oils.	78
6.9	Additive Reformulation for Bio Lubricants	79
6.10	Chemical Modification of Base Oils	79
6.11	Synthetic Base Oil Synthesis	82
6.12	Bio-degradable Lubricants.	83
6.13	Conclusion	83
	References	84

7   Development of Biodegradable Lubricants **85**
7.1	Introduction	85
7.2	Drive for Environmental Compatible Lubricating Fluids	86
7.3	Application of Bio-lubricants.	87
7.4	Vegetable-Based Bio-lubricants and the Environment.	90
7.5	Biodegradable Base Stocks and the Environment	92
7.6	Basic Eco-toxicological Properties of Bio-lubricants	94
7.7	Development of High-Performance Industrial Bio-lubricants	95
7.8	Development of Bio-lubricants Technical Properties.	95
7.9	Bio-lubricant Limitation: Additives and Modification Process	96
7.10	Bio-degradable Greases	98
7.11	Bio-lubricants Potential for Long-Term Use	98
7.12	Biodegradation Accelerants for Lubricants	99
7.13	Bio-based Lubricants Market and Potential	100
7.14	Conclusion	100
	References	101

8   Lubricant Life Cycle Assessment. **103**
8.1	Introduction	103
8.2	Petroleum Mineral Base Oil.	104
8.3	Synthetic Ester Base Oil.	104

8.4   Vegetable Base Oils 105
8.5   LCA for Lubricants 105
8.6   Mineral Base Lubricant LCA........................... 105
8.7   Synthetic Ester Lubricant LCA 107
8.8   Rapeseed Base Oil LCA............................... 108
8.9   Environmental Impact 109
8.10   Conclusion 112
References.. 113

9   Environment and the Economics of Long Drain Intervals 115
9.1   Introduction 115
9.2   Lubricant Consumption Control......................... 116
9.3   Extended Drain Interval 117
9.4   Technology to Enhance Bio-lubricants for Extended Drain 119
9.5   Bio-lubricant Base Oil Market.......................... 121
9.6   Transformer Insulating Fluids 122
9.7   Elevator Hydraulic Fluid 122
9.8   Other Hydraulic Fluids................................ 123
9.9   Metal Working Fluid 123
9.10   Chain Cutter Bar Oils................................. 123
9.11   Wire Rope Grease..................................... 124
9.12   Railroad Lubricants 124
9.13   Lubricant Condition Monitoring and Extended Drain......... 124
9.14   Conclusion 125
References.. 126

10   Recycling of Used Oil 127
10.1   Introduction 127
10.2   Used Oil Disposal Challenges 129
10.3   Waste Oil Recycling Basics 130
10.4   Acid-Clay Re-refining Process.......................... 132
10.5   Hylube Process....................................... 133
10.6   Mineralöl Raffinerie Dollbergen (MRD) Solvent
        Extraction Process Using N-Methyl-2-Pyrrolidone........... 135
10.7   Vaxon Process.. 137
10.8   CEP Process 138
10.9   Cyclon Process....................................... 140
10.10   Snamprogetti Process/IFP Technology..................... 140
10.11   Revivoil Process...................................... 141
10.12   Latest Used Oil Re-refining Technologies 142
10.13   Recycling of Waste Engine Oils Using Acetic Acids.......... 145
10.14   Conclusion 147
References.. 148

11 Environment and the Economics of Long Drain Interval 149
    11.1   Introduction 149
    11.2   Environmental Protection Aspects 152
    11.3   Application of Environmentally Friendly Lubricants 153
    11.4   Improved Lubricity. 156
    11.5   Need for Improved Oxidation Stability 156
    11.6   Viscosity-Temperature Behavior 157
    11.7   Evaporation Loss 158
    11.8   Bio-lubricant Cost Aspect Advantage 158
    11.9   Current EAL Markets. 158
    11.10 Bio-based Eco-friendly Lubricants of Recent Times 159
    11.11 Inadequacies of Current Bio-based Eco-friendly Lubricants. ... 160
    11.12 Future Bio-based Eco-friendly Lubricants. 160
    11.13 Conclusion 163
    References. 163

12 Environmentally Adapted Lubricants. 165
    12.1   Introduction 165
    12.2   New Lubricant Requirements. 167
    12.3   Base Oil Fluids for Environmentally Adapted
              Lubricants (EALs) 167
          12.3.1   Vegetables. 168
          12.3.2   Synthetic Esters 169
          12.3.3   Polyalkylene Glycols 169
          12.3.4   Biodegradability. 170
          12.3.5   Toxicity. 170
          12.3.6   Bio-accumulation 171
    12.4   Expectation on Additives 173
    12.5   Toxicity Levels. 173
          12.5.1   Acceptable Bio-degradability. 174
    12.6   Bio-accumulation of Lubricants. 175
    12.7   Eco-labeling Schemes and Regulatory Initiatives 175
    12.8   The Future of Environmentally Friendly Lubricants (EFLs). ... 176
    12.9   Conclusion 177
    References. 178

13 Proper Lubricants Handling 179
    13.1   Introduction 179
    13.2   Drive for Proper Oil Handling 181
    13.3   Possible Contamination in Storage 181
    13.4   Over Fill Protection and Containment 182
    13.5   Used Oil Handling 183
    13.6   Spill Protection. 184
    13.7   Waste Oil Recommended Disposal 184
    13.8   Environmental Fate of Lubricants 185

13.9   Lubricant Handling Recommendations at Operational Sites.... 186
13.10   Conclusion 187
References 187

14 Lubricating Grease Handling and Waste Management 189
14.1   Introduction 189
14.2   Lubricating Grease Structure Composition 190
14.3   Applications of Greases 190
14.4   Grease Manufacturing Process........................... 191
14.5   Environmental Compatibility of Greases.................. 191
14.6   Biodegradability of Greases............................. 192
14.7   Green Lubricating Greases.............................. 193
14.8   Base Oils for Lubricating Greases 194
14.9   Vegetable or Natural Base Oils 195
14.10   Chemically Modified Natural Base Oils 195
14.11   Biodegradable Synthetic Base Oils (Esters). 196
14.12   Thickening Agents 196
        14.12.1   Biodegradable Thickeners 196
        14.12.2   Non Biodegradable Environmentally Friendly
                  Clay Thickeners 198
        14.12.3   Synthetic Polymeric Thickeners................. 198
14.13   Additives in Lubricating Greases......................... 200
14.14   Lubricating Grease Challenges 200
14.15   Used or Waste Lubricating Greases 201
14.16   Impact on Human Species 202
14.17   Impact on Aquatic Species.............................. 202
14.18   Economic Impact 203
14.19   Waste Lubricating Grease Management.................... 203
        14.19.1   Collection and Handling........................ 203
        14.19.2   Transportation and Storage...................... 204
        14.19.3   Disposal 204
        14.19.4   Treatment 204
14.20   Conclusion 205
References 206

15 Beyond Lubricating Oil and Grease Systems................... 207
15.1   Introduction 207
15.2   Lubrication Systems.................................... 208
15.3   Lubrication Concept.................................... 209
15.4   Surface-Active Materials 209
15.5   Practical Lubricant Application 210
15.6   Lubricant Constituent Material 211
15.7   Operation of Solid Lubricants 212
15.8   Anti-seize Pastes....................................... 213
15.9   Anti-Friction Coatings (AFCs)........................... 214

15.10   Advantages and Limitations of Using AFCs   216
15.11   Typical Applications of AFCs   216
15.12   Drive for Solid Lubricants   217
15.13   Nano-materials and Limitations in Lubricant Application.   220
15.14   Conclusion   222
References   223

16   Conclusion   225
16.1   Future Lubricant Technology Drivers   225
16.2   Synthetic Engine Oil Revolution   225
16.3   Fate of Lubricant Additives   226
16.4   Environmental Friendly Lubricants Reality   228
16.5   Concluding Remarks   228
References   229

Appendix   231

Glossary   235

About the Authors

Ignatio Madanhire is a Ph.D. student with the School of Engineering Management at the University of Johannesburg, South Africa. He is holder of B.Sc. (Hon) in Mechanical Engineering and M.Sc. in Manufacturing Systems and Operations Research from the University of Zimbabwe. He has research interests in cleaner production, industrial optimization, production facility redesign, product life cycle assessment and operations management. The author has 20 peer-reviewed articles in international journals and conference proceedings, and Ignatio Madanhire has a quest for contributing to the industrial and engineering body of knowledge in the area of sustainable development. Ignatio Madanhire has served as a Mechanical Engineer for Government of Zimbabwe. He has also served as a Lubrication Engineer with Mobil Oil International and Lubricants Business Development Manager with Castrol International. He is currently a lecturer in Mechanical Engineering Department at the University of Zimbabwe. In addition, he is a consultant in industrial lubrication and manufacturing systems. He is a member of the Zimbabwe Institute of Engineers (ZIE).

Charles Mbohwa is the Vice Dean of Postgraduate Studies, Research and Innovation, Faculty of Engineering and the Built Environment, University of Johannesburg, South Africa. As an established researcher and professor in sustainability engineering and operations management, his specializations include renewable energy systems, bio-fuel feasibility and sustainability, life cycle assessment and healthcare operations management. He has presented at numerous conferences and published more than 150 papers in peer-reviewed journals and conferences, 6 book chapters and one book. Upon graduating with a B.Sc. in Mechanical Engineering from the University of Zimbabwe in 1986, he served as a Mechanical Engineer at the National Railways of Zimbabwe, Zimbabwe. He holds an M.Sc. in Operations Management and Manufacturing Systems from the University of Nottingham, UK and completed his doctoral studies at Tokyo Metropolitan Institute of Technology, Japan. Professor Mbohwa was a Fulbright Scholar visiting the Supply Chain and Logistics Institute at the School of Industrial and Systems Engineering, Georgia Institute of Technology. He has been a collaborator with the United Nations Environment Programme and a Visiting Exchange Professor at Universidade Tecnologica Federal do Parana. He has also visited many countries on research and training engagements including the UK, Japan, Germany, France, USA, Brazil, Sweden, Ghana, Nigeria, Kenya, Tanzania, Malawi, Mauritius, Austria, the Netherlands, Uganda, Namibia and Australia. He is a fellow of the Zimbabwe Institution of Engineers and a registered Engineer with the Engineering Council of Zimbabwe.

Abbreviations

ABT	Acute bacterial toxicity
ADO	Automotive diesel oil
AFC	Anti-friction coating
AFT	Acute fish toxicity
AOMT	Acute oral mammalian toxicity
API	American Petroleum Institute
ART	Auto-restoration technology
ATF	Automatic transmission fluid
BCF	Bio-concentration factor
EAL	Environmentally adapted lubricant
EC	Ethyl cellulose
ECL	Environmentally considerate lubricants
EFL	Environmentally friendly lubricant
EGR	Exhaust gas recirculation
EHL	Elastohydrodynamic viscosity index
EP	Extreme pressure
EVA	Ethylene-vinyl acetate
FTIR	Fourier transform infrared
GHG	Greenhouse gases
HD	Heavy duty
HDPE	High-density polyethylene
HMW	High molecular weight
HOB	High over base
HOSO	High-oleic sunflower oil
HOVO	High-oleic vegetable oil
HVI	High viscosity index
IL	Ionic liquid
LCA	Life cycle assessment
LDPE	Low-density polyethylene
LMW	Low molecular weight
LOB	Low over base

LPG	Liquefied petroleum gas
LTV	Limited threshold value
MEK	Methyl ethyl ketone
MSDS	Material safety data sheet
MWF	Metal working fluid
NMP	N-methyl-2-pyrrolidone
NPG	Neopentyl glycol
OEM	Original equipment manufacturer
OSHA	Occupational safety and hazard assessment
PAG	Polyalkylene glycol
PAH	Polyaromatic hydrocarbon
PAO	Polyalphaolefins
PCB	Polychlorinated biphenyls
PCMO	Passenger car motor oils
PE	Pentaerythritol
PEPE	Perfluoropolyester
pH	Acidity
PLC	Product life cycle
PM	Particulate matter
PP	Polypropylene
ppm	Parts per million
SAE	Society of Automotive Engineers
SAP	Sulphated ash and phosphorus
SAPS	Sulphated ash, phosphorus and sulphur
TAN	Total acid number
TDA	Thermal de-asphalting
TFE	Thin film evaporation
TMP	Trime thylol propane
VHVI	Very high viscosity index
VOC	Volatile organic compounds
WEN	Water endangering number
ZDTP	Zinc dialkyldithio phosphates

List of Figures

Figure 1.1   Typical hydrocarbon configuration. *Source* Wills (2005)   2
Figure 1.2   Crude processing fractional distillation products   3
Figure 1.3   Lubes oil processing activities. *Source* Madanhire
             and Mugwindiri (2013)   4
Figure 1.4   Typical blending yard product flow. *Source* Madanhire
             and Mugwindiri (2013)   4
Figure 1.5   Small pack filling line. *Source* Madanhire and
             Mugwindiri (2013).   5
Figure 1.6   Drum filling line. *Source* Madanhire and
             Mugwindiri (2013).   5
Figure 1.7   Stribeck lubrication diagram   8
Figure 1.8   Worldwide consumption of lubricants
             (Salimon et al. 2010)   9
Figure 1.9   Overview of the environmental impacts of regeneration
             versus the primary production   11
Figure 1.10 Lubricants product life cycle. *Source* Madanhire
             and Mugwindiri (2013)   13
Figure 2.1   Ashless succinimide dispersants   25
Figure 2.2   Calcium sulphonates   25
Figure 2.3   Zinc dialkyldithiophosphates.   26
Figure 2.4   Ashless inhibitors.   26
Figure 2.5   Development of lubricant additives (ATC 2007).   27
Figure 2.6   Component additive package formulation.   28
Figure 2.7   Flow diagram on fate of lubricants   30
Figure 2.8   Fate of lubricants   31
Figure 3.1   Contribution HD diesel emissions.   36
Figure 3.2   Potential reduction of sulphur   37
Figure 3.3   Noack volatilities of different base oil types versus
             specification limits.   39
Figure 3.4   Fuel economy benefits obtained from VHVI
             group III base oil   39

Figure 3.5   Typical truck particulate emissions   40
Figure 3.6   Composition of current commercial lubricants versus
              the assumed future limits.   41
Figure 3.7   Specific emissions of the mineral and synthetic engine oils. . . .   43
Figure 4.1   Life cycle of polymeric material based on vegetable oils
              (adapted from Samarth et al. 2015)   49
Figure 4.2   Thermal stability of additives   53
Figure 4.3   Typical PDSC 2-peak deposit method thermo-gram   54
Figure 5.1   Viscosity grade/base stocks on fuel economy   61
Figure 5.2   A generalized lubricant degradation diagram
              (Li et al. 2010)   62
Figure 5.3   Poly alphaolefin reaction   64
Figure 7.1   Some benefits of bio lubricants   90
Figure 8.1   Life cycle of mineral-based hydraulic fluid in a total
              loss application   106
Figure 8.2   Material flows for the production of a synthetic
              ester-based hydraulic fluid.   107
Figure 8.3   Formation of a typical synthetic ester, trimethylol
              propane trioleate (TMP trioleate)   108
Figure 8.4   Material flows for the production of rapeseed
              oil-based hydraulic fluid in a total loss application   109
Figure 8.5   The global warming potential contribution for
              the base fluids   111
Figure 8.6   The relative contribution to global warming
              potential of CO_2, CH_4 and N_2O for the base fluids   111
Figure 8.7   Global warming potential saved when 3 % less
              diesel is used for the base fluids   111
Figure 8.8   The total acidification potential of the base fluids.   112
Figure 9.1   Stress on passenger car lubricant.   117
Figure 9.2   Typical lubricant degradation pattern   118
Figure 9.3   Lubricant degradation with additive dose replenishment   118
Figure 9.4   Hydraulic oil—synthetic and mineral oil comparison.   119
Figure 10.1   Flow diagram of the typical used oil re-refining process.   131
Figure 10.2   Block flow diagram of the acid-clay re-refining process.   133
Figure 10.3   Block flow diagram of the Hylube process   134
Figure 10.4   Block flow diagram of the MRD Solvent extraction
              process.   136
Figure 10.5   Vaxon process block flow diagram   138
Figure 10.6   Block flow diagram of the CEP process   139
Figure 10.7   Flow diagram of the Cyclon process   140
Figure 10.8   Snamprogetti process block flow diagram.   141
Figure 10.9   Block flow diagram of the Revivoil process   142
Figure 10.10 A block flow diagram for solvent extraction   144
Figure 10.11 Flow diagram of hydro-finishing.   144
Figure 10.12 Worldwide lubricating oils consumption.   146

Figure 11.1   Diesel particulate emissions in relation to evaporation
              losses of oils (European 13-stage test, CEC R 49,
              Fuchs Steyr project).   155
Figure 11.2   Friction coefficients of base fluids—two disc test
              $\rho_H = 1000$ N/mm^2; $v_\Sigma = 8$ m/s; $\theta_{oil} = 60$ °C   156
Figure 11.3   Oxidative stability of biodegradable fluids   157
Figure 11.4   Viscosity-temperature behavior of base fluids
              (arrows indicate the differences of application
              temperature between an oil with VI = 100 and
              an ester based lubricant with VI = 200)   157
Figure 11.5   Evaporation loss of base fluids—IS0 VG 32.   158
Figure 11.6   Atomic structure of an ionic liquid (IL)   162
Figure 13.1   Application and disposal flow chart.   180
Figure 13.2   Lubricant spill containment set up.   183
Figure 15.1   Tribosystem showing parts in relative motion, lubricant
              film, load and operating environment   208
Figure 15.2   Stribeck curve for friction between surfaces relative
              to viscosity, speed and load   209
Figure 15.3   Suitable lubricant forms for different friction states
              (boundary, mixed and hydrodynamic regimes)   210
Figure 15.4   Operating mechanisms of solid lubricants.   213
Figure 16.1   Drivers of future oil formulations   226

List of Tables

Table 1.1	Typical additive proportion	6
Table 2.1	Effects of waste lubricating oil on physical properties of soil	22
Table 2.2	Effects of waste lubricating oil on the chemical properties of soil	22
Table 2.3	Relative substance toxicity	29
Table 3.1	API lubricant base stock categories	37
Table 3.2	Typical drain intervals over the recent past (km)	38
Table 3.3	Sources of SAPS (sulphated ash, phosphorous and sulphur) in lubricants (additives)	41
Table 3.4	Physico-chemical characteristics of mineral and synthetic base oils	42
Table 4.1	Oil contents of widely available non edible vegetable oils	48
Table 4.2	The biodegradability of petroleum based lubricants and lubricants from alternative resources	50
Table 4.3	Comparison of petroleum based lubricants with lubricants from alternative resources	51
Table 7.1	Common terms related to the environment	93
Table 7.2	Benefits of biodegradable lubricants	94
Table 8.1	Weighting factors used to calculate the global warming potential (time horizon 20 years)	110
Table 8.2	Weighting factors for the acidification potential	110
Table 9.1	Engine oil use trends	116
Table 10.1	Properties of used oils, intermediate products during re-refining	132
Table 10.2	API base oil categories	132
Table 10.3	Properties of base oil products of Hylube-process	135
Table 10.4	Properties of base oil products of MRD solvent extraction process	137
Table 10.5	Properties of base oil products of CEP process	139
Table 10.6	Properties of base oil products of Revivoil process	142
Table 11.1	Lubricants in the environment (for Germany)	150

Table 11.2 ISO 15380:2002 requirements for environmentally
            acceptable hydraulic fluids based on synthetic esters 150
Table 11.3 Typical base fluids for EAL against mineral oil base 151
Table 11.4 Total-loss lubricants currently available 153
Table 11.5 Emission reduction levels of modern engine oils 155
Table 12.1 Differential biodegradation rates by lubricant base oils 170
Table 12.2 Comparative toxicity of base oils 171
Table 12.3 Bioaccumulation potential by base oil types. 172
Table 12.4 Water hazard classification 174
Table 15.1 Lubricant constituent components for specific
            regime requirements 211
Table 15.2 Types of anti-seize pastes by color, solid-lubricant
            content and typical applications 214
Table 15.3 Common solid lubricants and characteristics 218
Table 15.4 Areas where solid lubricants are required. 219
Table 15.5 Advantages and disadvantages of solid lubricants 219

Chapter 1
Introduction

Abstract The crude processing is put into perspective as the source of mineral base to make lubricants. The base oil and additives after physical mixing constitute the lubricating oils. Two main categories of lubes are automotive, marine, aviation and industrial. Uncontrolled disposal of lubricant has adverse effect on the soils, aquatic life and renders water unfit for drinking. Current life cycle emphasizes used oil disposal as the critical phase of the lube to be paid greater attention to reduce environmental impact.

1.1 Background

The documented detrimental effect of petro-chemicals on flora and fauna, demands that close monitoring is done for the entire process of manufacturing, application and disposal to safeguard the environment. Hence need to investigate this area and recommend possible areas of improvement to comply with emerging and demanding regulatory requirements. Although considerable research has been devoted to impact of petroleum fuels, a rather less attention has been paid to pollution resulting from lubricants on disposal. It would seem, therefore, that further investigation is needed to explore the lubricant products impact on the environment from manufacturing to disposal. It is imperative that an investigation is done to mitigate the lubricants' impact on the environment in line with the ever tightening regulatory compliance in this regard.

A growing number of companies involved with commercial marketing of lubricants are faced with a mounting challenge of mitigating the impact of lubricants on the environment during and after use. It has been observed that a greater number of efforts have been directed at dealing with environmentally friendly disposal of used oil in the product life cycle (PLC) given in Fig. 1.9. This approach overlooks the equally detrimental effect of fresh oil during usage and the blending process to the environmental. The common impacts of lubricants if disposed off without due care are soil degradation and water contamination resulting, in some instances, in loss aquatic life.

© Springer International Publishing Switzerland 2016
I. Madanhire and C. Mbohwa, *Mitigating Environmental Impact of Petroleum Lubricants*, DOI 10.1007/978-3-319-31358-0_1

Manufacturers have a duty to do everything they can to minimize the impact of our products on the environment and on people, directly and indirectly. One way to do this is to replace non-renewable components with ingredients that are renewable and biodegradable. Another way is to develop new products that perform better than the existing ones. When lubricants contribute to vehicles, machines and factories being more energy efficient, less energy is used. And reducing friction is where the greatest environmental benefits can be gained.

It is in this regard that deliberate effort is undertaken to reduce negative environmental impacts of lubricants throughout the value chain from blending to disposal by manufacturers and users. As well as establishing gaps in the current handling of lubricants to improve on the relevant practices. Such an initiative would entail issues on base oil substitution, minimizing oil losses during use, economic extended oil drain intervals, used oil recycling and effective oil disposal ways. Lubricant producers are held responsible for the lube disposal. This could be easily done if manufacturers collaborate with end users for sustainable development (Madanhire and Mugwindiri 2013).

1.2 Mineral Base Oil Processing

Crude petroleum is a complex mixture of hydrocarbons. While crudes from different sources vary in chemical composition, they all possess the same basic components. Light hydrocarbon gases, gasoline, diesel fuel, and fuel oil fractions are distilled from the crude, leaving residuum, from which mineral lubricating oil stocks are derived. A variety of sophisticated refining techniques are used for the removal or beneficial transformation of undesirable components such as asphalts, waxes, and sulphur and nitrogen compounds.

What remains is a lubricating oil base stock that, despite the degree of chemical refinement, still contains a complex mixture of organic compounds as shown in Fig. 1.1, including oxygen, sulphur, and nitrogen compounds, plus a variety

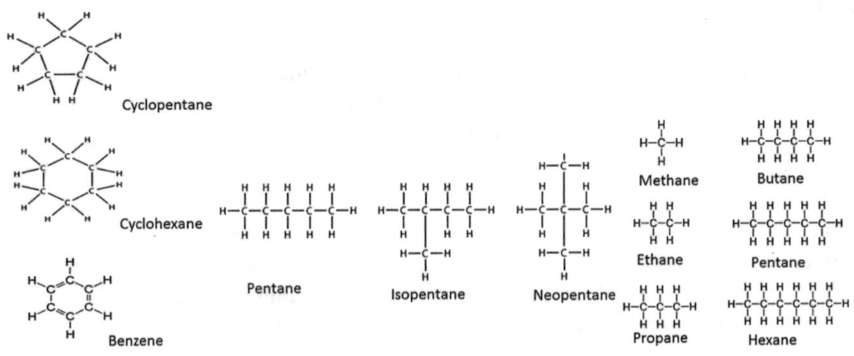

Fig. 1.1 Typical hydrocarbon configuration. *Source* Wills (2005)

of cyclic hydrocarbons, as well as small amounts of inorganic substances. Of the thousands of molecular structures present in crude oil, many, or most, of them are removed by refining; however, many other different structures will remain in mineral lubricating oil stocks, whether manufactured by solvent—or hydrogen processing techniques (Murphy et al. 2002). The boiling ranges of the compounds increase with the number of atoms as given here:

- Far below −18 °C for light natural gas hydrocarbons with one to three carbon atoms
- About 204–343 °C for gasoline components
- 400–650 °C for diesel and home heating oils
- Higher ranges for **lubricating oils** and heavier fuels.

Crudes range from "paraffinic" types which are high in paraffinic hydrocarbons, through "intermediate" or "mixed" to the naphthenic types, which are high in hydrocarbon containing ring structures.

Various fractions are tapped out of the crude processing as shown in Fig. 1.2, and of interest is the mineral base oil, which is the major constituent of liquid lubricants. By virtue of it being a high boiling point fraction, it means it has high thermal stability and can withstand high temperatures of 650 °C in a diesel vehicle engine.

The mineral base oil from the crude processing is further processed by removing the undesirable materials from the final product using various processes. These processes include vacuum distillation, where the base oil feedstock is separated into products with similar boiling point range. This is followed by the de-asphalting process were asphalt related compounds are removed. The third process is called the furfural extraction, where the removal of aromatic compounds is done to get de-asphalted raffinate. Methyl ethyl ketone (MEK) de-waxing is the fourth process, where de-waxed oil is produced. This free of wax oil base oil is used in certain premium application. The processes are finalized by hydro-finishing, a catalytic reaction with hydrogen to decolorize the base oil and resulting in chemically stable base oil. Thus this series of processes ensures that crude oil used in lubricating oil manufacture is good base oil and free of undesirable material (Wills 2005).

Fig. 1.2 Crude processing fractional distillation products

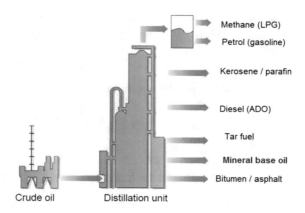

1.3 Lubricants Blending

The process of coming up with a mineral liquid lubricant is divided into main processes which are upstream operation and downstream operation. Where upstream involves exploration, mining and crude distillation refining to get lube base, whereas downstream entails raw material sourcing, storage, lube blending, packaging and distribution as shown in Fig. 1.3.

Most non crude producing countries are involved only with the downstream operation of physically mixing the raw materials in a lube blending plant as given by Fig. 1.4. Most synthetic base oil range originates from Europe and Asia because of the technology involved. All the additives used are commercially available with from suppliers such as Lubrizol (Madanhire and Mugwindiri 2013).

Lubricants are composed of a majority of base oil and a minority of additives to impart desirable characteristics indicated. Common groups of base oil are 150SN (solvent neutral), 500SN, and 150BS (bright stock).

Although generally lubricants are based on one type of base oil or another, it is quite possible to use mixtures of the base oils to meet performance requirements.

Production usually takes place in 1000–30,000 L batches, which are analyzed by the quality control laboratory prior to being transferred into holding tanks, followed by packaging into small packages and drums as shown in Figs. 1.5 and 1.6. This ensures that the finished product meets the set quality requirements.

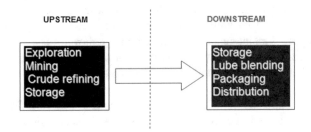

Fig. 1.3 Lubes oil processing activities. *Source* Madanhire and Mugwindiri (2013)

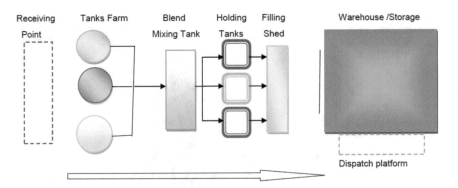

Fig. 1.4 Typical blending yard product flow. *Source* Madanhire and Mugwindiri (2013)

Fig. 1.5 Small pack filling line. *Source* Madanhire and Mugwindiri (2013)

Fig. 1.6 Drum filling line. *Source* Madanhire and Mugwindiri (2013)

In summary the three major functions involved in the lube blending process sequence in Fig. 1.4 are: Receiving and storage of base oils, additive and packaging materials; physical mixing as per formulation sheet and agitation blending of base oil and additives, and filling of packaging containers with oil; and warehouse storage and distribution of finished lubricants.

1.4 Lubricants Types and Applications

According to Madanhire and Mugwindiri (2013) a **lubricant** (sometimes referred to as "lube") is a liquid substance introduced between two moving metal surfaces to reduce the friction between them, improving efficiency and reducing wear.

A lubricant's ability to lubricate moving parts and reduce friction is the property known as lubricity. The single largest application for lubricants is protecting the internal combustion engines in vehicles as motor oil.

Typically lubricants contain 90 % **base oil** (or mineral oils) and less than 10 % **additives**. Also synthetic liquids such as hydrogenated poly olefins, esters, silicones and fluorocarbons are sometimes used as base oils. Additives deliver reduced friction and wear, increased viscosity, improved viscosity index, resistance to corrosion and oxidation, aging and contamination among others (Wills 2005).

$$\text{LUBRICANT} = \text{BASE OIL} + \text{ADDITIVES}$$
$$\qquad\qquad\qquad 90\,\%\qquad\qquad 10\,\%$$

A large number of additives are used to impart performance characteristics to the lubricants. The main families of additives which constitute the 10 % composition are:

- Ant oxidants
- Detergents
- Anti-wear
- Metal deactivator
- Corrosion inhibitor/Rust inhibitor
- Friction modifiers
- Extreme pressure (EP)
- Anti-foaming agent
- Viscosity index improver
- Demulsifying/emulsifying
- Stickiness improver
- Complexing agents for greases.

Many of the basic chemical compounds used as detergents (example: calcium sulfonate) serve the purpose of the first seven items in the list as well. Usually it is not economically or technically feasible to use a single do-it-all additive compound. Oils for hypoid gear lubrication will contain high content of EP additives. Grease lubricants may contain large amount of solid particle friction modifiers, such as graphite, molybdenum sulfide etc.

Typical additive proportions can be summarized as below for various applications as required is given in Table 1.1.

Table 1.1 Typical additive proportion

Lubricant application	Percentage (%)
Transformer	1
Hydraulic	5
Gear oil	8
Cutting oil	10
Engine oil	15

Types of lubricants are mostly based on their applications are categorized by usage as: Automotive (petrol and diesel), Aviation (turbine and piston), Marine, and Industrial (hydraulics, compressors, gears, bearing circulation systems, turbines and refrigeration systems). At micro level the main functions of the lubricants as given below (Madanhire and Mugwindiri 2013):

Keep moving parts apart: Lubricants are typically used to separate moving parts in a system. This has the benefit of reducing friction and surface fatigue together with reduced heat generation, operating noise and vibrations. Lubricants achieve this by forming a physical barrier i.e., a thin layer of lubricant separates the moving parts.

Reduce friction: The lubricant-to-surface friction is much less than surface-to-surface friction in a system without any lubrication. Thus use of a lubricant reduces the overall system friction. Reduced friction has the benefit of reducing heat generation and reduced formation of wear particles as well as improved efficiency. Lubricants may contain additives known as friction modifiers that chemically bind to metal surfaces to reduce surface friction even when there is insufficient bulk lubricant present for hydrodynamic lubrication, e.g., protecting the valve train in a car engine at startup.

Transfer heat: Liquid lubricants are much more effective on account of their high specific heat capacity. Typically the liquid lubricant is constantly circulated to and from a cooler part of the system, although lubricants may be used to warm as well as to cool when a regulated temperature is required. This circulating flow also determines the amount of heat that is carried away in any given unit of time.

Carry away contaminants and debris: Lubricant circulation systems have the benefit of carrying away internally generated debris and external contaminants that get introduced into the system to a filter where they can be removed. Lubricants for machines that regularly generate debris or contaminants such as automotive engines typically contain detergent and dispersant additives to assist in debris and contaminant transport to the filter and removal.

Transmit power: Hydraulic fluids are used as the working fluid in hydrostatic power transmission. Hydraulic fluids comprise a large portion of all lubricants produced in the world. The automatic transmission's torque converter is another important application for power transmission with lubricants.

Protect against wear: Lubricants prevent wear by keeping the moving parts apart. Lubricants may also contain anti-wear or extreme pressure additives to boost their performance against wear and fatigue.

Prevent corrosion: Good quality lubricants are typically formulated with additives that form chemical bonds with surfaces to prevent corrosion and rust.

Seal for gases: Lubricants will occupy the clearance between moving parts through the capillary force, thus sealing the clearance. This effect can be used to seal pistons and shafts.

Also lubricants must provide a liquid seal at moving contacts and remove of wear particles. In order to perform these roles, lubricating oils must have specific physical and chemical characteristics. Perhaps the fundamental requirement of

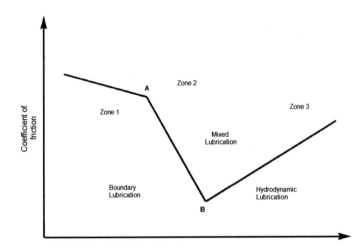

Fig. 1.7 Stribeck lubrication diagram

lubricants is that the oil should remain a liquid over a broad range of temperatures. In practice, the usable liquid range is limited by the pour point (PP) at low temperatures and the flash point at high temperatures. The PP should be low to ensure that the lubricant is pump-able when the equipment is started from extremely low temperatures. The flash point should be high to allow the safe operation and minimum volatilization at the maximum operating temperature. For the most demanding applications, such as aviation jet engine lubricants, an effective liquid range over 300 °C may be required. The efficiency of the lubricant in reducing friction and wear is greatly influenced by its viscosity. The relationship between speed, viscosity, load, oil film thickness, and friction is illustrated by the Stribeck diagram in Fig. 1.7.

The coefficient of friction for bearing is plotted against the dimensionless duty parameter $\mu N/F$, where μ is the dynamic viscosity of the lubricant, N the rotational speed of the shaft, and F is the loading force per unit area.

1.5 Impact on Environment

Unlike petroleum fuel which is burnt and emitted as gaseous by-products to the environment, for which exhaust-system catalytic converters have since been fitted to vehicles to release less harmful gases. Also low emission drive by engine manufacturers means the more harmful substances escape into the lubricants in engine oil application. Thus on disposal, used oil loaded with these is normally dumped as liquid waste. This poses a serious environmental damage in form of soil degradation, water contamination and interference with ecosystem balance.

Lubricants both fresh and used can cause considerable damage to the environment mainly due to their high potential of serious water pollution. The additives contained in lubricant can be toxic to flora and fauna. In used fluids the oxidation products can be highly toxic as well. Lubricant persistence in the environment largely depends upon the base fluid, however if very toxic additives are used, they cause this negative persistence.

Impact of lubricants on the environment places primary emphasis on the used lubricants since they may contain materials that are harmful to life or the environment. Issues of lubricant conservation, used oil reclamation, reprocessing, disposal and oil biodegradability are being taken up by a number of researchers to mitigate adverse effects of oil on the environment. Approximately 37.9 million metric tons of lubricants were used globally in 2005, with a projected increase during the next decade at 1.2 %. The majority of these lubricants are typically petroleum-based. Various environmental reasons have prompted the exploration of alternative lubricants, including synthetic and bio-based lubricants. Vegetable oils are potential substitutes for petroleum-based oils because they are environmentally friendly, renewable, less toxic and readily biodegradable.

The challenge remains the ever-increasing use of lubricants as shown by Fig. 1.8 by various regions by the world. Thus extended service intervals and recycle of used lubricant have been widely adopted to deal with this menace. In order to attain the extended service interval, one must use lubricants with extended useful life. Recycling is the option to minimize the used lubricants entry into the environment. This translates into cost savings, with respect to buying a batch of a new lubricant as well as in disposal costs, and the potential damage to the environment, if the disposal method is inappropriate. To minimize inadvertent entry of the lubricant into the environment, a closed and self- contained system of used oil collectors in workshops are installed and connected to the used oil storage tank with pipes and a diaphragm pump. This has in some cases successfully minimized the loss of the lubricant or its volatile components into the environment through leakage and evaporation (Madanhire and Mugwindiri 2013).

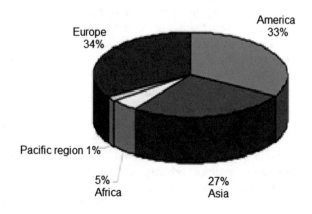

Fig. 1.8 Worldwide consumption of lubricants

1.6 Lubricants Handling

Recycling, burning, landfill and discharge into sewage and waterways are some of the disposal methods observed for used lubricant. There are strict regulations in most countries regarding disposal in landfill and discharge into water as even small amount of lubricant can contaminate a large amount of water. Most regulations permit a threshold level of lubricant that could be present in waste streams and companies spend hundreds of millions of dollars annually in treating their waste waters to get to acceptable levels.

Burning the lubricant as fuel is also governed by regulations mainly on account of the relatively high level of additives present. Burning generates both airborne pollutants and ash rich in toxic materials, mainly heavy metal compounds. Thus lubricant burning takes place in specialized facilities that have incorporated special scrubbers to remove airborne pollutants and have access to landfill sites with permits to handle the toxic ash.

Unfortunately, most lubricant that ends up directly in the environment is due to general public discharging it onto the ground, into drains and directly into landfills as trash. Other direct contamination sources include runoff from roadways, accidental spillages, natural or man-made disasters and pipeline leakages. Occasionally, unused lubricant requires disposal. The best course of action in such situations is to return it to the manufacturer where it can be processed as a part of fresh batches. Improper handling of lubricants can cause soil and water contamination. It can also damage water drinkability, recreational use and fisheries. Recent German research has revealed that as little as one liter of mineral oil can render one million liters of drinking water unfit for human consumption.

While there are many don'ts and preventive measures that can be recommended to people who are handling used oils in auto repair shops, foremost among the measures should be to dispel these workers', as well as self-servicing vehicle owners', pathetic attitude. The authors observed that while there is a common warning found in the labels of engine oil containers, seldom do mechanics or users in general, if ever, read the important information contained therein and take necessary precautions, i.e., to avoid exposure to and dispose of properly the used synthetic or mineral oils.

To avoid being subjected to the harmful ingredients in used oils, some precautions can be undertaken. There are many oil-resistant gloves created for the purpose and these should be worn all the time while working with engines to prevent the direct effects of used oil. There should be used oil tanks made available for lubricant disposal or collection. This will require, therefore, an effective campaign among mechanics in auto repair shops and general users to handle used oils carefully and inform them of ways to dispose of these hazardous substances properly.

In countries with regeneration process capability, re-refining is the most recommended way to go. The major environmental impact of waste oil regeneration is to relieve the burden of primary production of lubricants as shown by Fig. 1.9. It was established that regeneration brings about important net relief with respect to six environmental performance indicators which are: resource depletion, greenhouse

Fig. 1.9 Overview of the environmental impacts of regeneration versus the primary production

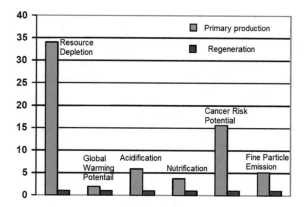

effect, acidification, nutrification, carcinogenic risk potential and fine particle emissions.

Figure 1.9, illustrates the ratio between the average impact of regeneration in relation to the aforementioned indicators, and the equivalent burden that would result from primary production. It demonstrates that regeneration has a considerably lower environmental burden than the processes it substitutes. It can be seen that the lower the impact the better for the environment. With respect to the other recovery options where waste oils are used as a heavy or light fuel, regeneration outperforms incineration from an ecological perspective in relation to the six environmental performance indicators. Furthermore with the evolution towards an increasing use of synthetic base fluids in lubricants, the omitted environmental burden resulting from regeneration substituting primary production will increase in the future (GEIR 2011).

1.7 Disposal and Harmful Ingredients

It is estimated that less than 45 % of used engine oil is being collected while the remaining 55 % is thrown by the end user into the environment. Auto repair shops are among the highest contributors of improper used oil disposal (on internet http://www.articlesbase.com/automotive-articles/used-oil-and-its-effects-on-the-environment-876009.htm).

Auto repair shops are a common feature in most countries. As high mileage cars expectedly do not work efficiently due to parts that are at the verge of their usefulness, this requires frequent visits to the mechanic. Among those common maintenance works performed by car or small vehicle repair shops are engine overhauling and oil change. Engines have to be re-bored and sleeved to bring the car's piston to its original size. For both old and new cars, regular replacement of degraded oil has to be done to keep the engine running smoothly.

As a result of these maintenance works, waste oil disposal by these small repair shops becomes a concern. Waste oil disposal is taken lightly by many users and handlers partly due to their ignorance of the ill effects of improper disposal and partly due to lack of clear environmental guidelines and weak enforcement of environmental laws aimed towards the management of oil wastes. Mechanics in repair shops or the self-servicing vehicle owner just dispose of their wastes directly into the drain, unmindful of the possible human and environmental consequences.

Karachi University in Pakistan analyzed used motor oil samples from across their city. They found out that the average level of concentration of different metals found in the samples was:

> **lead** 110 ppm (parts per million),
>
> **zinc** 685 ppm,
>
> **barium** 18.1 ppm,
>
> **arsenic** 5 ppm,
>
> **cadmium** 2.5 ppm and **chromium** 3.2 ppm.

The soil contaminated with used motor oil had about:

> 100 ppm **arsenic**,
>
> 20 ppm **cadmium**,
>
> 1800 ppm **lead** and 285 ppm **barium**.

The relatively high concentration of metals in contaminated soil is due to accumulation of these metals. The above metals are harmful ingredients found in many synthetic oils. All these metals are highly toxic with carcinogenic and teratogenic effects.

Carcinogenic substances are substances that can cause cancer because of their ability to damage the cell or interfere with its normal metabolic processes. **Teratogenic substances** are substances that can interfere with normal embryonic development. Exposure is through direct contact and through ingestion via the food chain (on internet http://www.highbeam.com/doc/1G1-104970182.html). This means that the harmful metal ingredients contained in used oils can increase the incidence of cancer among humans as well as lead to increased cases of abnormal babies. Continued exposure to these substances can, therefore, be dangerous as these can directly and indirectly affect people (Madanhire and Mugwindiri 2013).

1.8 Biodegradability and Toxicity

Biodegradability and toxicity have been combined because they both pertain to similar characteristics of lubricants. The degree of biodegradability of a lubricant relates to the environmental friendliness of a lubricant, whereas the toxicity of a lubricant relates to the friendliness of the lubricant to the user. Biodegradability is becoming a more important property as environmental regulations become more restrictive with regard to protecting the integrity of water supplies and the environment in general. Toxicity has been an important issue for quite some time, and safety to the user will continue to gain attention. So far, the most important aspect of a lubricant for a specific application is that it work well, and the environmental friendliness is secondary. In the future, it could be that the environmental aspects of the lubricant will take precedence over performance (Gschwender et al. 2001).

1.9 Lubricant Life Cycle

"Lube life cycle" refers to the notion that a fair, holistic assessment requires the assessment of base oil production, manufacture, distribution, use and disposal including all intervening distribution steps necessary or caused by the lubricant's existence. The sum of all those steps—or phases—is the life cycle as indicated by Fig. 1.10. The concept also can be used to optimize the environmental performance of lubricants to optimize the environmental performance of an organization.

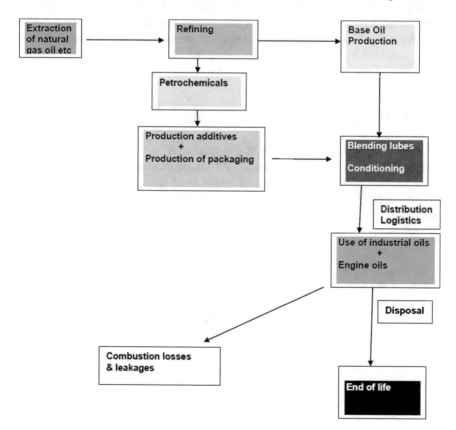

Fig. 1.10 Lubricants product life cycle. *Source* Madanhire and Mugwindiri (2013)

Life cycle analysis methodology is used for identifying and reducing the impact of lubricant on the environment. With this approach, respect for the environment is built into the new product design right from the start. From initial design to disposal of a lubricant, life cycle analysis enables examining and monitoring environment impacts arising from lubricant using ISO 14040 standard procedures. One can assess environmental impact and power consumption of each stage in the life cycle of the lube: production, utilization and end-of-life. The main goal is to identify key points in the cycle requiring most attention with aim of reducing environmental impacts at the very most. By investigating environment impacts of lubricant, it has been identified that end-of-cycle is the key stage for problems which occur in handling and disposal of used mineral based lubricants.

Effective lubricant lifecycle management begins with the careful decision to select a lubricant that possesses the performance properties required for the application, the operating context and the operating environment. By managing temperature, moisture, particles and other contaminants, the life of the lubricant can be proactively extended and oil analysis can provide the necessary information to make the oil change decision, assuming the application is a suitable candidate for condition-based drains. No matter how diligent one is in the management of lubricant life, all lubricants will at some point require changing. Be certain the drain effectively eliminates the used oil and that the used oil is properly disposed of in accordance with applicable regulations.

1.10 Conclusion

The background to understanding of the origins of petroleum mineral base oil is key in the unpacking of their impact on the environment. It is also important for readers to appreciate the constituent make up of the lubricants as base oil and additives for the preceding chapters which dwell on the detailed detriment of the product to the environment.

Revision Questions

1. *How does mineral base oil processing ensure quality lubricant performance?*
2. *What is the average proportion of additives in lubricants?*
3. *Which are the major categories of lubricants?*
4. *List key additives which are found in lubricants*
5. *What criteria are used to drain the lubricant from the equipment?*

References

GEIR—Groupement Européen de l'Industrie de la Régénération (2011) An environmental review of waste oils regeneration: why the regeneration of waste oils must remain an EU policy priority, The Re-refining Industry Section of UEIL, Brussels, Belgium

Gschwender LJ, Kramer DC, Lok BK, Sharma SK, Snyder CE Jr, Sztenderowi ML (2001) Liquid lubricants and lubrication. In: Modern tribology handbook. CRC Press LLC (2001)

Madanhire I, Mugwindiri K (2013) Cleaner production in downstream lubricants industry. Lambert Academic Publishing, Saarbrucken, Germany

Murphy WR, Blain DA, Galiano-Roth AS (2002) Benefits of synthetic lubricants in industrial applications. ExxonMobil Research and Engineering Company, Paulsboro, New Jersey, USA

Wills JG (2005) Lubrication fundamentals. Marcel Dekker Inc., New York, USA

Chapter 2
Lubricant Additive Impacts on Human Health and the Environment

Abstract It is estimated that, at present, approximately 50 % of all lubricants sold worldwide end up in the environment via total loss applications, volatility, spills or accidents. More than 95 % of these materials are mineral oil based. In view of their high eco-toxicity and low biodegradability, mineral oil-based lubricants make up a considerable threat to the environment. While, most lubricants and hydraulic fluids based on plant oils are rapidly and completely biodegradable and are of low eco-toxicity; moreover, lubricants based on plant oils display excellent tribological properties and generally have very high viscosity indices and flashpoints. However, in order to compete with mineral-oil-based products, some of their inherent disadvantages must be corrected, such as their sensitivity to hydrolysis and oxidative attack, and their behavior at low temperatures. The chapter also makes effort to characterize the potential human health and environmental hazards of widely used classes of lubricating oil additives and poly-alpha-olefin (PAO) base fluids, and the related toxicity levels.

2.1 Introduction

Original equipment manufacturers (OEM) specifications for lubricant performance will continue to drive changes in formulations, with particular focus on lubricants' contribution to meeting fuel economy and emissions regulations leading to increased demands placed on related lubricant characteristics. Environmental concerns will continue to play a major role in lubricant formulation and use. Reduction of elements such as chlorine, phosphorus, sulfur, and metals has proceeded at a rapid pace over the past decade, particularly in automotive lubricants. Use of more environmentally friendly fuels, including renewable fuels, in both automotive and industrial engines will also drive changes in lubricant formulation and additive demand. For example, expanding use of biodiesel in the motor vehicle fuel pool will require better oxidation and corrosion protection from lubricants. In contrast, falling sulfur content in marine fuel oil may reduce the need for detergents in marine engine lubricants (Freedonia 2013).

© Springer International Publishing Switzerland 2016
I. Madanhire and C. Mbohwa, *Mitigating Environmental Impact of Petroleum Lubricants*, DOI 10.1007/978-3-319-31358-0_2

A number of ways to improve the undesirable properties of native plant oils are being pursued by various researchers. While governments are putting regulations in place to enforce the use of bio-based fluids, for use in ecologically sensitive areas. Here effort is made to look at the key impact of additives on human health and the environment. The main classes of additives are: succinimide ashless dispersants, calcium sulphonates, calcium phenates, zinc dialkyldithiophosphates, oxidation inhibitors, and anti-wear inhibitors. Although lubricant additives do not pose a significant health risk to humans, lubricant additives do not readily biodegrade and may persistent in the environment. Manufacturing advances are reducing the release of toxic by-products to the environment. The potential health effects of lubricant additives in humans can be determined in appropriate animal toxicity tests. This testing is an important tool to communicate accurately the human health and environmental hazards through the use of appropriate labels, Material Safety Data Sheets (MSDS), and employee training.

2.2 Environmental Effects of Used Oil

When the lubricants are drained from say engines, gearboxes, hydraulic systems, turbines and air compressors: the oil is contaminated with wear debris; the lubricating base oil has deteriorated and degraded to acids; the additives have decomposed into other chemical species; and process fluids such as degreasers and solvents have mixed into the used oil. It was also noted that used oil contains wear metals such as iron, tin and copper as well as lead from leaded petrol used by motorists. Zinc arises from the additive packages in lubricating oils. Many organic molecules arise from the breakdown of additives and base oils. The molecule potentially the most harmful is the polycyclic aromatic hydrocarbon (PAH) such as benz(a)pyrene and chrysene. Petrol engines generate the most PAH molecules per 1000 km, with diesel engines below that and two-stroke engines generating the least amount of PAHs.

Any release of used oil to the environment, by accident or otherwise, threatens ground soil and surface waters with oil contamination there by endangering drinking water supply and aquatic organisms. Used oil has been established that it can damage the environment in several different ways such as:

- Spilled oil tends to accumulate in the environment, causing soil and water pollution. Oil decomposes very slowly. It reduces the oxygen supply to the microorganisms that break the oil down into non-hazardous compounds.
- Toxic gases and harmful metallic dust particles are produced by the ordinary combustion of used oil. The high concentration of metal ions, lead, zinc, chromium and copper in used oil can be toxic to ecological systems and to human health if they are emitted from the exhaust stack of uncontrolled burners and furnaces.

- Some of the additives used in lubricants can contaminate the environment e.g., zinc dialkyl dithiophosphates, molybdenum disulphide, and other organo-metallic compounds.
- Certain compounds in used oil—e.g., poly-aromatic hydrocarbons (PAHs)—can be very dangerous to one's health. Some are carcinogenic and mutagenic. The PAH content of engine oil increases with operating time, because the PAH formed during combustion in petrol engines accumulates in the oil.
- Lubricating oil is transformed by the high temperatures and stress of an engine's operation. This results in oxidation, nitration, cracking of polymers and decomposition of organ-metallic compounds.
- Other contaminants also accumulate in oil during use—fuel, antifreeze/coolant, water, wear metals, metal oxides and combustion products.

Thus in summary, in summary if used oil is disposed in an irresponsible manner it may cause great danger to the human resources like water supply. The environmental effects of used oil can be classified as human health effects, wetlands and wildlife effects, burning waste effects, marine and fresh water organisms effects, and effects of using waste oil as dust control.

2.3 Environmental Impacts

An important cause of soil biodiversity loss in urban areas is the pollution of soils by petroleum products. Engine lubricating oil is a major product of petroleum which helps the engine move smoothly. Spent or waste engine oil is oil that has been used, and as a result contaminated by chemical impurities which contribute to chronic hazards including mutagenicity and carcinogenicity as well as environmental hazards with global ramifications. Waste engine lubricating oil is oil that has served its service properties in a vehicle withdrawn from the meant area of application and considered not fit for initial purpose. Waste engine oil is a mixture of several different chemicals including low and high molecular weight aliphatic hydrocarbons, aromatic hydrocarbons, polychlorinated biphenyls, chloro dibenzofurans, lubricative additives, decomposition products and heavy metal contaminants such as aluminum, chromium, lead, manganese, nickel and silicon that come from engine parts as they wear down.

Waste engine oil has also been shown to create an unsatisfactory condition for life in the soil. At the impacted site arthropod assemblage and abundance was much lower, suggesting that spent lubricating oil from car engines demonstrated profound effect on the arthropod species. The absence of myriapod species as well as the generally low individual numbers of other arthropods at the contaminated sites may be explained by habitat transformation which resulted in biodiversity loss and elimination of species in the habitat. At the impacted site arthropod assemblage and abundance was much lower, suggesting that spent lubricating oil from car engines demonstrated profound effect on the arthropod species.

The absence of myriapod species as well as the generally low individual numbers of other arthropods at the contaminated sites may be explained by habitat transformation which resulted in biodiversity loss and elimination of species in the habitat (Rotimi and Ekperusi 2014).

According to literature reports, varying amounts of lubricants at present, end up in the environment. For example, heavy duty marine engines require both a cylinder lubricant and a crankcase lubricant. The cylinder lubricant requires a very high level of detergency to cope with acids produced in the combustion of lower quality marine fuels, while water separation and deposit control are most important for the crankcase oil. Natural gas engines contend with higher temperatures and high levels of nitrogen oxides (known as nitration). Sulfur containing EP additives can be further described by their content of "active" sulfur. Active sulfur refers to the tendency of sulfur to react chemically at low temperatures. High active sulfur content is desirable when EP performance is needed at comparatively low temperature, but can also lead to corrosion, particularly of copper. Extreme pressure additives are produced by a number of companies, and sold as part of additive packages as well as individual additive components (Freedonia 2013).

There are many different modes of action through which an oil or lubricant may adversely affect species and these include: direct toxicity of oil or lubricant components, especially to the liver and kidneys in mammals and fish. And indirect toxicity by reducing the breeding success and by passage of compounds into the young or through suppression of the immune system. This may be through bioaccumulation or bio-concentration of compounds over a long period of time.

In the USA 32 % (432×10^6 gallons of 1351×10^6 gallons) of lubricating oils ended up in landfills or were dumped. It is claimed that 50 % of all lubricants sold worldwide end up in the environment via total loss, spillage and volatility. Estimates for the loss of hydraulic fluids are as high as 70–80 %. Most problematic are uncontrolled losses via broken hydraulic hoses or accidents whereby large quantities of fluids escape into the environment. They contaminate soil, surface, ground- and drinking water and also the air.

While, lubricants and hydraulic fluids based on plant oils are generally rapidly and completely biodegradable and are also of low eco-toxicity. At present the use of pure native plant oils is limited to total loss applications (lubricants for chainsaws, concrete mould release oils) and those with very low thermal stress. Hydraulic fluids are of increasing importance for applications in environmentally sensitive areas where a potential total loss could be encountered, such as excavators, earthmoving equipment and tractors, in agricultural and forestry applications and in fresh water (groundwater) sensitive areas. Mineral oils are toxic for mammals, fish and bacteria. Considering the sump capacities of such machinery (up to 1000 l) the ecological impact is obvious—as are the economics of the resulting clean up operations.

Although it seems obvious that the increased use of rapidly biodegradable lubricants would be of considerable ecological and economical advantage, the present market share of these materials is relatively small. For hydraulic fluids this amount is increasing more rapidly, with estimates ranging from 25 to 75 %.

Technical performance, acceptable price and ecological compatibility will constitute the basis for future developments along these lines. From some studies, it seems that bio lubricants are already available for the majority of applications and that the technical performance is comparable and sometimes even better than for conventional lubricants. As they display:

- excellent tribological properties (ester functions stick well to metal surfaces)
- lower friction coefficients than mineral-oil-based fluids
- lower evaporation (Noack)—up to 20 % less than mineral-oil-based fluids
- higher viscosity index (multi-range oils)
- excellent biodegradability
- high flashpoints
- low water pollution classification.

Their technical properties are thus largely comparable with mineral-oil-based fluids. However, they are thermally less stable than mineral oils, sensitive to hydrolysis and oxidative attack, and their low-temperature behavior is frequently unsatisfactory. For the development of lubricants and hydraulic fluids on the basis of renewable resources such as native plant oils one will always have to make a compromise between the performance based on the chemical structure and the desired biodegradability and eco-toxicity. All lubricants and hydraulic fluids are composed of so-called base fluids and additives.

Major arguments for environmentally acceptable plant-oil-based lubricants and hydraulic fluids are (a) high biodegradability and (b) low eco-toxicity. Yet in spite of these criteria the term environmentally acceptable lubricants must be properly defined, especially for formulated oils. A formulated oil consisting of a highly biodegradable base fluid in combination with toxic additives environmentally is not acceptable: while the bulk of fluid is indeed biodegradable, the overall lubricant is not. The lower the degree of un-saturation, the longer is the lifetime. Saturated synthetic esters are clearly the most stable ester materials at present. However, in many cases the biodegradability is less than satisfactory. At present it seems that the best compromise between performance, price and biodegradability are high oleic oils. These can be obtained either by cultivation (selective breeding), with high oleic sunflower oil (90 % + oleic acid) being the most prominent example, followed in the future possibly by high oleic rapeseed and soybean(selective breeding or GMO). Alternatively, chemical modifications of commodities such as rapeseed oil or biodiesel would be an interesting alternative.

2.4 Effect of Used Oil on Soil

Waste lubricating oil having been contaminated with impurities in the course of usage and handling, contain toxic and harmful substances such as benzene, lead, cadmium, polycyclic aromatic hydrocarbons (PAHs), zinc, arsenic, polychlorinated biphenyls (PCBs) etc. which are hazardous and detrimental to the soil and the surrounding environment. The environment must be protected against pollution by lubricants and hydraulic fluids based on mineral oils. This is, of course, best done by preventing undesirable losses and by reclaiming and reusing lubricants. Alternatively, environmentally acceptable lubricants and hydraulic fluids should be used whenever and wherever possible. Increase in demand for cars, heavy duty automobiles, generators etc. throughout the year, led to increase in demand for lubricating oils, and this resulted in the generation of large oil irrespective of the type and source of collection, is sometimes dumped into the soil causing harmful or toxic materials to percolate through the soil thus contaminating the soil and thereby changing the physical and chemical properties. It is also sometimes dumped down drain, sewers, disrupting the operations at waste water treatment plants (Udonne and Onwuma 2014).

Sample analyses were done on two samples and the results indicated that the presence of lubricating oil in one of the samples had altered the soil chemistry, and thus resulted in the adverse effects on the physical and chemical properties of the soil. Table 2.1 shows how the physical properties of were altered, when looking at properties like bulk density, capillarity, porosity, and water holding capacity.

While on a separate set of samples, the presence of waste oil had an equally adverse effects on the chemical properties of the soil as shown on Table 2.2. Waste lubricants reduced the pH thereby making the soil more acidic. There was considerable reduction in potassium content in the contaminated soil, as well as increase in carbon contents this being attributed to carbon present in waste oil. An increase in moisture was also noted, as originated from used oil.

Thus it was amply shown by experiments that waste lubricating oils can significantly and adversely affect physical and chemical properties of the soil making it less productive, due to the various contaminants within the used

Table 2.1 Effects of waste lubricating oil on physical properties of soil

Parameters	Uncontaminated soil sample 'A'	Contaminated soil sample 'B'
Bulk density (g/cm^3)	1.10	1.15
Soil capilarity (cm/h)	8.10	0.04
Soil porosity (ml)	110	80
Water holding capacity (WHC) (ml)	55.0	15.0

Table 2.2 Effects of waste lubricating oil on the chemical properties of soil

Parameters	Uncontaminated soil sample 'A'	Contaminated soil sample 'B'
Soil pH	6.5	6.0
Phosphorus content (ppm)	80	40
Potassium content (ppm)	98	60
Organic carbon	2.15	3.05
Moisture content (%)	3.5	9.9

lubricants such as OCBs, PAHs, benzene, lead, arsenic, zinc etc. which are the cause for soil contamination unlike virgin oil.

2.5 Future Development of Eco-friendly Lubricants

In general terms, the cost of biodegradable lubricants and hydraulic fluids is— depending on the product quality—frequently higher than that of comparable mineral-oil-based products. Since the price level is also dependent on the corresponding additivation and the amounts purchased. differences. Generally, the price of rapeseed oil is much lower than that of synthetic esters, with differences of up to 200 %. However, the price differences have to be seen also in relation to differences in quality and extended lifetime (i.e., extended periods between oil changes). The current differences in prices between bio-based and mineral-oil-based fluids may, however, become irrelevant if one looks into the very near future. It can be predicted that mineral oil production will reach its peak. Since bio-lubricants are already available for numerous applications, with performance being well comparable and sometimes even better than those of mineral oil products—combined with the methods for chemical derivatizations increasingly being developed—plant oils are an attractive alternative as raw materials for future lubricants.

2.6 Chemistry of Lubricant Additives and Their Toxicology to Humans

Neither mineral nor synthetic base oils can satisfy today's lubricant performance requirements without using additives. Additives are chemical substances, in most part synthetic, which are used in lubricant formulations to adjust a broad of spectrum of properties by enhancing what is desired and suppressing what is unwanted. Many additives are multifunctional products that may exhibit synergistic or antagonistic behavior when mixed together. As a rule of thumb, additives do not add. This makes balancing and optimization of additive systems a challenging task.

The increasing focus on energy efficiency and environmental safety of lubricants poses new challenges for lubricant formulators, preventing or restricting the use of certain time-proven chemistries, such as ZDDP in engine oil or boric acid in MWF formulations. At the same time, it stimulates the search for new classes of additives, including all-organic ashless friction modifiers, nano-additives, and bio-based super lubricity additives, as well as fundamental studies into how individual additives work.

Exposure to lubricating oil additives usually occurs by skin contact with finished oils that contain the additives. On rare occasions, workers may be exposed to the neat additive in the manufacturing and blending plants or during transportation. The criteria for identifying the hazards of petroleum additives are based on those of the EU Dangerous Substances Directive or the US OSHA Hazard Communication Standard. The risks are normally classified as below:

- Acute oral toxicity
- Acute dermal toxicity
- Eye irritation
- Skin irritation
- Skin sensitization.

The interpretation of test results for other toxicity end points, such as genotoxicity, reproductive and developmental toxicity, and target organ effects, requires the evaluation of expert toxicologists to determine the potential hazards.

Ashless succinimide dispersants: The most widely used and manufactured lubricating oil additives are ashless succinimide dispersants (Fig. 2.1). They are used at levels approaching 50 % in all automotive crankcase additive packages. They are also found in marine, railroad diesel, natural gas-fuelled engine and air- and water-cooled two-cycle engine lubricants. Frequently, additive suppliers will functionalise succinimides with additional materials, such as boric acid (borated succinimides) and organic acids (oxalic, mono short-chain alkyl, terephthalic, etc.).

Toxicity of succinimide ashless dispersants: They are not considered harmful if swallowed or absorbed through the skin. They are neither eye nor skin irritants, and they do not cause an allergic skin reaction based on guinea-pig sensitization tests. Unlike mono succinimides and borated succinimides, which gave negative results in the Ames, succinimides functionalized with organic compounds are positive.

Sulphonates and phenates: Sulphonates and phenates are important members of the group of additives known as metallic detergents. Sulphonates are used in virtually all types of lubricants, while phenates are used primarily in marine, railroad, and automotive crankcase oils. Sulphonates are prepared either from the sulphonation of lubricant oil (natural sulphonates) or from the sulphonation of synthetically prepared C_{16} and greater alkylbenzenes (synthetic sulphonates). Sulphonates can be prepared as calcium, magnesium, or, less importantly, barium and sodium salts. Sulphonates are prepared in order to produce either low over based (LOB) or high overbased (HOB) products (Fig. 2.2). The basic metallic carbonate over basing is present in HOB sulphonates as a reverse micelle (Fig. 2.2).

Toxicity of sulphonates and phenates: Calcium sulphonates are generally not hazardous but there is some concern with skin sensitization potential of calcium sulphates. These materials are not considered harmful if swallowed or absorbed through the skin. They are neither eye nor skin irritants. They do not cause serious systemic effects in subchronic toxicity tests.

Zinc dialkyldithiophosphates: Zinc dialkyldithiophosphates (ZDTPs) are the most popular multifunctional additives used (Fig. 2.3). They act as anti-oxidants, anti-wear agents, and corrosion inhibitors. ZDTPs are made by the reaction of alcohols, phosphorus pentasulphide (P_2S_5), and zinc oxide. The type of alcohol

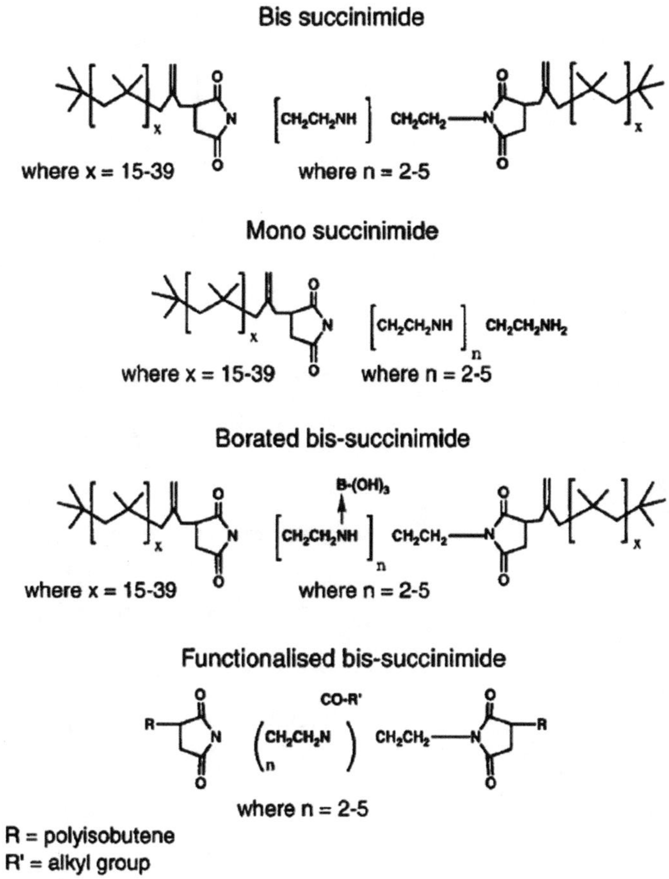

Bis succinimide

where x = 15-39

where n = 2-5

Mono succinimide

where x = 15-39

where n = 2-5

Borated bis-succinimide

B-(OH)$_3$

where x = 15-39

where n = 2-5

Functionalised bis-succinimide

CO-R'

where n = 2-5

R = polyisobutene
R' = alkyl group

Fig. 2.1 Ashless succinimide dispersants

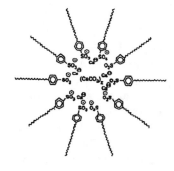

Low-overbased calcium sulphonate

$R = C_{16}$ plus

High-overbased calcium sulphonate

$R = C_{16}$ plus

High-overbased calcium sulphonate reverse micelle

Fig. 2.2 Calcium sulphonates

Fig. 2.3 Zinc
dialkyldithiophosphates

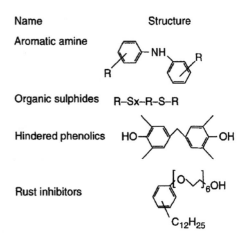

used to make a ZDTP can vary. Low molecular weight C_{3-6} secondary alcohols are used to prepare ZDTPs for use in automotive lubricants. These additives decompose at low temperatures, thus acting as very good wear inhibitors. Industrial oil applications use ZDTPs made from primary C_{4-8} alcohols. In these types of ZDTPs, oxidation performance is enhanced relative to secondary ZDTPs.

Toxicity of Zinc dialkyldithiophosphates: The primary human health hazard of concern with ZDTPs is their eye irritation potential though not irritating to the skin. It was found out that ZDTPs are mutagenic. Further work showed that the mutagenicity of ZDTPs was due to the presence of zinc.

Rust inhibitors: Rust inhibitors such as the one shown in Fig. 2.4 are used in a variety of industrial and crankcase lubricants. Organic fatty acids made from a variety of natural and synthetic sources are also used extensively in the lubricant industry as rust inhibitors.

Toxicity of rust inhibitors: They are not expected to be harmful if swallowed or absorbed through the skin. They are neither eye nor skin irritants, and they do not cause allergic skin reactions. The molybdenum-nitrogen complex is not mutagenic, and it does not cause serious systemic effects in subchronic toxicity tests.

Products such as demulsifiers, friction modifiers, antifoam agents, pour point depressants, and viscosity index improvers are combined with the additive classes discussed above to formulate performance additive packages which are sold to lubricant manufacturers around the world. Lubricant manufacturers formulate finished lubricant products using refinery produced lubricant base stocks, PAOs, and other synthetically derived base stocks.

2.7 Lubricant Additives and Their Hazards to the Environment

As given in Fig. 2.5 below gives a chronological view of the development of the main additive families since 1930s to present. These developments have been driven by new specification demands imposed by engine design changes, which in turn are a response to consumer demand and emissions requirements. Base oils or synthetic base stocks alone cannot provide all the engine lubricant functions required by a modem gasoline or diesel engine. Over the last eighty years a number of chemical additives have been developed to enhance base stock properties, overcome their deficiencies and provide the new performance levels required by the technological evolution of engines or by new regulations (ATC 2007). Lubricant additives fall into two categories:

- those protecting metal surfaces in the engine, such as anti-wear, anti-rust, anticorrosion and friction modifier additives; and,
- those reinforcing base stock performance, such as antioxidants, dispersants, viscosity modifiers and pour point depressants.

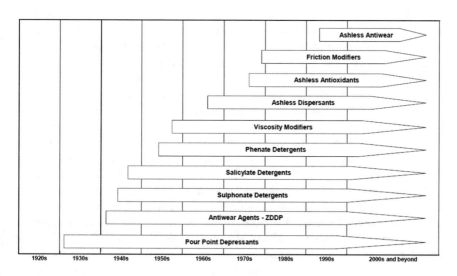

Fig. 2.5 Development of lubricant additives (ATC 2007)

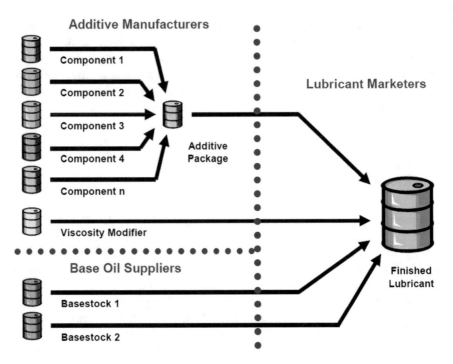

Fig. 2.6 Component additive package formulation

Finished lubricants contain a number of individual additive components—typically about eight but ranging from five to fifteen. Some or all may be blended individually into the lubricant base stock during manufacture. More typically, the components are pre-blended by the additive manufacturer into a performance additive package which is sold to the lubricant marketer. The viscosity modifier, which is a major component of multi-grade crankcase lubricants, is usually purchased and blended separately by the lubricant marketer. Figure 2.6 illustrates this schematically.

The majority of lubricant additives are of low mammalian toxicity and are typically less harmful when ingested than familiar household products. Some lubricant additives are suspected of being harmful to aquatic organisms and most show a degree of persistence but these additives are typically of low water solubility and when handled and disposed of according to manufacturers' recommendations are considered not to present a significant environmental risk.

The composition of the crankcase lubricant changes during use. Some additives are chemically changed or even destroyed as part of their functionality, and any contaminants of combustion generated during the course of engine operation which are not swept into the exhaust stream are neutralized and dispersed within the lubricant. It is widely accepted that used oil drained from the engine sump will contain poly aromatic hydrocarbons (PAHs) generated by the combustion process and this waste is suspected to pose a carcinogen risk through accidental skin contact. Users are normally advised against inappropriate contact or disposal of used lubricants to minimize the environmental and human risks from used oil.

Lubricant additives can be released into the environment during manufacture, transport, and usage, and therefore it is necessary to evaluate these materials for their potential in causing adverse environmental effects. We use a standard approach to environmental hazard characterization that evaluates lubricant additives based on the following three criteria of environmental effects:

- the material's toxicity to sensitive environmental organisms,
- the material's relative persistence in the environment measured by its degradation rate,
- the material's potential to accumulate in the food chain.

Though all materials are evaluated for their potential in causing toxicity in both aquatic and terrestrial environments, actual toxicity testing is usually performed on aquatic organisms because they show greater sensitivity to materials than do terrestrial organisms. This is due to the fact that aquatic organisms are closely associated with their aqueous environment, and they are usually exposed to much higher concentrations of released materials for longer periods of time. Further, more adequate toxicity assessments can be made on terrestrial organisms by extrapolating the mammalian toxicity data for the material.

The following aquatic toxicity thresholds are generally recognized for classifying and labeling materials worldwide based on the amount of material required to produce adverse effects on representative aquatic test organisms (fish, aquatic invertebrates, and algae) when additives are directly added to the test water (Table 2.3).

When a lubricant additive is directly added to water, the test organisms are exposed to different fractions of the additive. These include the water-soluble portion, portions that form dispersions or emulsions, and the insoluble fraction which lies on the surface of the water or coats the container. Therefore, the effects observed on the organisms could be influenced by the non-soluble fractions of the material which may cause fouling (physical coating due to contact with the test organisms). Fouling effects may interfere with the observation of toxic effects and

Table 2.3 Relative substance toxicity

Relative substance toxicity	EC_{50}^{*} range (mg/l)
Relatively harmless	>1000
Practically non-toxic	>100–1000
Slightly toxic	>10.0–100
Moderately toxic	>1.0–10.0
Highly toxic	<1.0

$^{*}EC_{50}$ is defined as the concentration of material that would affect 50 % of the test organisms during the exposure period. The greater the EC_{50} the lower the toxicity of the material can be assessed by reviewing the physical/chemical properties of the lubricant additives.

These data show that lubricant additives are practically non-toxic to aquatic test organisms, except for ZDTPs and hindered phenolics. ZDTPs are moderately toxic to aquatic organisms. Hindered phenolics are highly toxic, but are used in finished oils at use levels of less than 1 %. The activated sludge inhibition test results indicate that none of the lubricant additives that were tested would disrupt the operation of waste water treatment facilities or adversely affect micro organisms in the environment.

2.8 Ultimate Fate of Lubricant Additives

All lubricants, as identified in the previous section are here followed through to their ultimate fate in the air, water or soil compartments. Other industry partners have placed increasing emphasis on the reuse and safe disposal of used oil. OEMs have given much emphasis over the last five years to the recycling of used engine oil. The re-refining of used oils has also increased.

In above diagram (Fig. 2.7) effort was made to trace the fate of crankcase lubricants from sale through to the environmental compartments. The environmental pathways for lubricants used in the passenger car and truck sectors are the same. Thus crankcase lubricant additives also follow these same general pathways, to their ultimate fate in the air, water or soil compartments. Crankcase

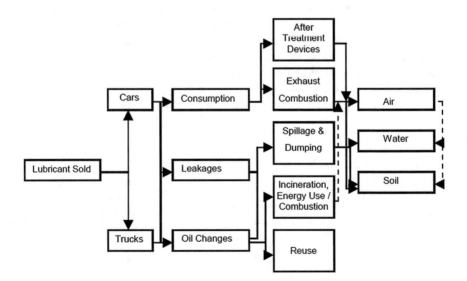

Fig. 2.7 Flow diagram on fate of lubricants

lubricants enter the engine when either the sump is filled or when the engine is topped up. They leave the engine:

- when drained, as used oil or
- as leakage through seals or gaskets during use or
- via lubricant consumption down the tailpipe as gaseous emissions (combustion products or particulates).

If the vehicle is fitted with a particulate trap, the incombustible portion of the lubricant in the exhaust will be trapped by this filter. Broken lines in Fig. 2.7 show that a portion of used oil which is collected and burnt or incinerated contributes to emissions in the air compartment—and that, ultimately, some emissions to the air compartment end in the soil or water compartments. The lubricant leaves the engine down the tailpipe as emissions of either combustion products (gases or vapor) or particulates (which will be collected if a particulate trap is fitted). A portion of the lubricant (and fuel) combustion products is entrained and re-dispersed in the crankcase lubricant itself, or forms deposits in the engine. Leakage past seals or gaskets is estimated to be negligible for both cars and trucks. The environmental routes of this portion of the lubricant are to the water and soil compartments.

Used oil should be drained from the crankcase, collected and re-refined and re-used, or disposed of suitably. Approximately 75 % of lubricants sold (2000 k tons) is drained as used oil at oil changes. This includes oil from scrap vehicles. A significant proportion (estimated 47 %) of that collected is used as fuel oil or incinerated for energy value or disposal. Regeneration represents an estimated 24 % of the total waste oil. Additives in oils properly disposed of by these routes end up as solids for landfill, used in cement or as gaseous emissions. However, approximately 28 % of used oil is still unaccounted for and may be dumped into the soil and water compartments, or burnt entering the air compartment. The proportions going to the various pathways are illustrated in Fig. 2.8.

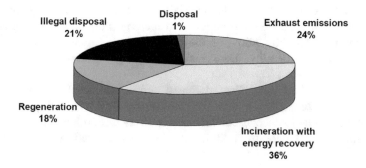

Fig. 2.8 Fate of lubricants

2.9 Biodegradation of Additives

If lubricant additives (or finished lubricants) are not consumed by combustion in an engine or recycled, we must concern ourselves with environmental fate. One possible measure of environmental fate is biodegradation. It is an important factor for determining the fate of lubricant additives in the environment together with water solubility, volatility, absorption to suspended solids, sediment and soil, hydrolysis, and photolysis. Although the biodegradation process is really a series of microbial-mediated processes which involve many kinds of microorganisms acting in concert to degrade both the materials and their degradation by-products, none of the current ready biodegradability tests adequately measures the entire process. In any of these tests, only the initial and terminal stages of the process are measured, i.e., primary and ultimate biodegradation.

Ultimate biodegradation, or loss of parent material from the test media through conversion into carbon dioxide and water (i.e., the extent of material mineralization), is determined in ready biodegradation tests that measure either carbon dioxide evolution from the test media or oxygen utilization (consumption) in the test media. In these tests, the criteria for designating materials 'readily biodegradable' is that more than 60 % of the carbon in the material has evolved as carbon dioxide in the test system, or that more than 60 % of the theoretical oxygen demand has been consumed in the test media, at the end of the 21–28 day tests.

For the majority of lubricant additives, degradation rates are usually very low because they have very limited water solubility, usually less than 1 mg/l. This limits their bioavailability to microorganisms and increases the length of time in which they will degrade. In addition, their very large molecular weights make them tend to partition from water to solids and organic matter where they are not readily accessible to microbial degradation processes. Currently the EU is concerned that slowly degradable materials may cause long-term harm to the environment due to their persistence. But a material's persistence in the environment is only an indicator of its fate, and not its effect on the environment. It has yet to be determined if rapidly or slowly degraded materials are really more environmentally desirable for a given use or disposal of a material. Rapidly degradable materials may deplete the available dissolved oxygen in the immediate environment and thereby indirectly affecting aquatic organisms through suffocation, or they may degrade rapidly into by-products that may be considerably more toxic to aquatic species than the parent materials.

There is a clear need for continued efforts to develop more relevant biodegradation tests that assess the biodegradability of lubricant additives under realistic environmental conditions as well as to develop better techniques of interpreting the data from these tests to provide realistic assessments of environmental effects of lubricant additives.

2.10 Bio-concentration of Additives

Materials that accumulate in the food chain to the extent they can cause adverse environmental and human health effects are cause for concern. The extent to which materials accumulate in organisms is expressed as the bio-concentration factor (BCF). In aquatic environments, BCF is defined as the ratio of the concentration of the material in the aquatic organism at equilibrium to the concentration of the material in the water. Generally, the less water soluble a material is, the greater its accumulation in the lipid tissues of aquatic organisms.

However, these tests were developed for materials consisting of single defined chemicals, not for materials like lubricant additives which are actually mixtures with a range of log Kow. Thus it is technically difficult to measure analytically all of the components of lubricant additives in fish bio-concentration tests and generally we rely on marker compounds in the lubricant-additive matrix. The testing of lubricant additives is further complicated by the fact that many components of lubricant additives are quite similar to naturally produced materials within the fish, thus interpretation of tissue residue results can be extremely difficult.

2.11 The Future of Additive Technology

The performance of synthetic sulphonates has matched the performance of natural sulphonates for most applications, but with the corresponding environmental benefits. An effort has been under way to replace diesel fuel and kerosene used to prepare drilling muds with fluids that may be less harmful. PAOs offer an advantage because they show significant biodegradation using the CEC L-33-T-82 biodegradation method, are stable under basic and acidic drilling conditions, and are much cheaper than more expensive esters and ethers.

Lubricant additive technology seeks to enhance performance of base fluids derived from a variety of crude oils. These lubricants can then be used with a wide variety of fuels derived from different crude oils and processes. Optimum use of petroleum resources is thus achieved. Lubricant additives are engine design components, enabling the continuing evolution of engine design to provide increasingly efficient and more environmentally-friendly vehicles. Lubricant additives provide high performance engineering for the consumer and ensure reliability, longevity and optimal performance of the vehicle. Lubricant additives are considered to be of low environmental risk, however there is need for ongoing health and safety reviews.

2.12 Conclusion

Considerable effort was spent to understand the potential human health and environmental hazards of lubricating oil additives, and these efforts have been described in this chapter. Future efforts will involve continuing to meet the human health and eco-toxicity requirements of new technology and meeting, to the fullest extent possible, the product stewardship and responsible care initiatives of industrial organizations and global regulatory agencies. There will be a special focus on understanding bio-accumulation and biodegradation of lubricants in the environment as our industry meets the challenges of the twenty-first century.

Revision Questions

1. *State three major hazards of additives on the environment*
2. *Name four classes of additive chemicals and explain their potential toxicity to humans.*
3. *What do understand by bio-degradation?*
4. *What physical nature of additives limits their bio-degradability?*
5. *Explain what is meant by bio-concentration effect of additives on aquatic life.*

References

ATC—Technical Committee of Petroleum Additive Manufacturers in Europe (2007) Lubricant additives and the environment, ATC Document 49 (revision 1), December 2007
Freedonia (2013) New US industry forecasts for 2017 & 2022: lubricant additives, industry market research for business leaders, strategists, decision makers, April 2013, www.freedonia-group.com
Rotimi J, Ekperusi OA (2014) Effect of spent lubricating oil on the composition and abundance of arthropod communities of an urban soil, J Sci Env Manage 18(3):411–416, www.bioline.org.br/ja
Udonne JD, Onwuma HD (2014) A study of the effect of waste lubricating on the physical/chemical properties of soil and the possible remedies, J Petrol Gas Eng 5(1) 9–14 (Academic Journals). http://www.academicjournals.org/JPGE

Chapter 3
The Environment and Lubricant Related Emissions

Abstract This chapter looks at the impact of lubricants on the environment including diesel exhaust emissions. Investigations have shown a clear effect of lubricant oil on emissions, which depends on lube oil characteristics, especially sulfur content, metal content, volatility and density. Engine lubricants help to improve vehicle efficiency but contribute engine exhaust emissions. More attention will have to be paid to off-road vehicles, especially tractors, if the production of healthy food and the maintenance of a cleaner environment are not to be compromised. Therefore, one the biggest challenges facing the automotive industry is to improve fuel economy, both to conserve natural resources and to limit pollutants and CO_2 emissions. Better fuel efficiency and consequently lower emissions will require new materials, new lubricants and low-emission fuel.

3.1 Introduction

The amount of emissions by automobile engines is also very strongly related to their fuel economy. In general, the higher the fuel economy the lower the emission of pollutant and carbon dioxide. One way to decrease emissions is to improve the fuel economy of vehicles. Rising fuel costs and the need to conserve fossil fuel have led to increased interest in the role of lubricants in improving fuel economy. Appropriate lubricant formulations can bring about a beneficial reduction in engine friction, thus improving fuel economy. Friction losses in a car engine may account for more than 10 % of the total fuel energy. The fuel consumption of gasoline and heavy duty diesel engines is of great importance, since it accounts for up to 30 % of operating costs.

Increased use of exhaust after-treatment technologies to meet tougher emissions legislation has proved a major challenge in the development of heavy-duty diesel engine lubricants that maintain extended oil drain intervals and wear protection. New regulations, especially for heavy-duty diesel engines, with ever stricter emission limits notably for particulate matter (PM) and nitrogen oxides (NO_x), are

© Springer International Publishing Switzerland 2016   35
I. Madanhire and C. Mbohwa, *Mitigating Environmental Impact of Petroleum Lubricants*, DOI 10.1007/978-3-319-31358-0_3

Fig. 3.1 Contribution HD
diesel emissions

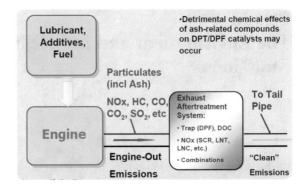

Fig. 3.1 Contribution HD diesel emissions

being introduced. As shown in Table 3.1, it is intended that in these regions NO_x and PM limits for diesel vehicles will be more than 80 % lower than today's levels.

The regulations are designed to reduce levels of particulates and NOx emissions from vehicle exhausts, due to their detrimental effect on air quality. In the past, OEMs have used a range of engine design measures to accommodate the emissions legislation, including modified combustion chambers and the use of EGR (exhaust gas recirculation). Having 'exhausted' the design tweaks under their control, the engine manufacturers have turned to exhaust after-treatment technology in order to comply. Since these after-treatment technologies are considered sensitive to the chemical composition of engine lubricants, this had major implications for the way lubricants are formulated and manufactured (Fig. 3.1).

A typical engine lubricant incorporates a base oil and a package of additives that may include viscosity improvers, anti-wear compounds, friction modifiers, corrosion inhibitors, dispersants, anti-foaming agents, anti-oxidants and detergents. Together, through their chemical compositions, they contribute to the so-called SAPS (Sulphated Ash, Phosphorus and Sulphur) profile of the lubricant. While SAPS have no direct effect on emissions, they have been associated with a negative impact on the performance and longevity of the new after-treatment hardware. The severity of the emissions standards requires the performance of after-treatment systems to be sustained over the life of the vehicle. As a result, lubricants for use after-treatment systems have had to be reformulated to lower the SAPS content. By using their extensive formulation expertise, and careful selection of the base oil and additive components, they have managed to meet the stringent new SAPS limits without compromising engine integrity. The reduction in sulphur as given in Fig. 3.2 has been achieved largely through the use of more sophisticated base oils. Traditional mineral base oils have been replaced with more highly refined varieties and synthetics (ATIEL 2006).

Emissions legislation continues to increase in severity in the automotive industry and will in the future be applied to an even greater degree to vehicles in the off-highway market. Attention will have to be paid to non-road diesel mobile machinery emissions, especially tractors, if the production of healthy food and

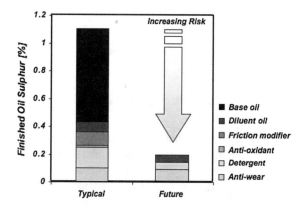

Fig. 3.2 Potential reduction of sulphur

the maintenance of a cleaner environment are not to be compromised. Non-road diesel engines account for about 47 % of diesel PM and about 25 % of total NO$_x$ emissions from mobile sources. Engine design changes required to meet the latest emission regulations greatly impact on the engine oil degradation process and, consequently, for each new emission regulation, a new engine oil specification is released.

From Chap. 1, it was noted that petroleum-based lubricant base oil is complex mixture of hydrocarbons with carbon numbers generally in the C$_{20}$ to C$_{40}$ range, depending on specific viscosity grade. Also the base oil contains a variety of molecules, such as paraffins, isoparaffins and naphthenes. American Petroleum Institute (API) lubricant base stock categories are shown in Table 3.1 below.

The chemical composition of base oil directly affects its performance. Compared with lubricant based mineral oils, synthetic lubricants have greater film thickness at high temperature, and achieve full lubrication more quickly at low temperature. Synthetic lubricants generally have better oxidative stability than comparable mineral-based lubricants. In future, automotive lubricants will need to be able to demonstrate improvements in fuel economy, extended drain intervals and reduced emissions.

Table 3.1 API lubricant base stock categories. *Source* Madanhire and Mugwindiri (2013)

API group	% Saturates	% Aromatics	% Sulfur	Viscosity index	Noack %
I	<90	>10	>0.03	<120	30
II	≥90	<10	≤0.03	80–120	25
III	>90	<10	<0.03	>120	11
IV		All polyalphaolefins			11
V		All stock not included in group I–IV			<11

Table 3.2 Typical drain intervals over the recent past (km)

Typical oil drain	1995	2003
Passenger car	10,000	30,000
Heavy-duty	40,000	100,000

3.2 Extended Drain Interval

Against the background of maintaining vehicle performance there has also been a strong trend towards extended drain intervals.

Table 3.2 demonstrates the significant changes in typical drain intervals in the recent past. Extending oil drains increases soot levels. If soot is not adequately dispersed by the engine oil, it can cause sludge to form on rocker and front engine covers, bearings to fail, valve bridges and fuel injection links to wear, and filters to plug. The durability of the lubricant and additive system in relation to the ability to disperse soot and maintain a regime of reduced wear has led to significant changes in additive formulation. Extending oil drains allows maximum up-time for hauling freight and reduces costs for fresh oil, filters, mechanics' labor, and used oil and filter disposal.

3.3 Fuel Economy Aspect

Original equipment manufacturers are under intense pressure to provide the most fuel-efficient vehicles to respond to both the emission regulations and Kyoto protocol. Already, several automobile manufacturers have applied very low viscosity 0W20 oils to their factory fills. These low-viscosity fuel-efficient oils are usually formulated with organic molybdenum compounds. They can provide a fuel economy benefit of as much as 2–3 %. In addition to the to the molybdenum friction-reducing additives, high-quality group III base oils also play a crucial role in the 0W20 oils. It has been pointed out that the use of low-viscosity oils could raise concerns about viscosity increases and oil consumption, which could negatively affect fuel efficiency and catalyst performance.

From this viewpoint, it was required that oil volatility must be reduced to 15 % Noack as shown in Fig. 3.3. This change not only minimizes the viscosity increase but also is expected to significantly reduce oil consumption. For low-viscosity 0W20 oils, this can be achieved only through the use of very high viscosity index group III base oils.

Figure 3.4 shows fuel economy benefits as a result of the lower elasto-hydrodynamic lubrication (EHL) friction of the very high viscosity index (VHVI) group III base oils in comparison with group I. Typically, polyalphaolefin (PAO) based engine oils have a fuel consumption benefit of up to 3.4 % relative to comparable mineral oils. In automotive transmissions the benefit is of the order of 10 % of the power transmitted through the unit, resulting in a fuel economy

Fig. 3.3 Noack volatilities of different base oil types versus specification limits

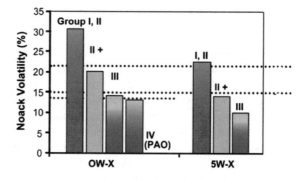

Fig. 3.4 Fuel economy benefits obtained from VHVI group III base oil

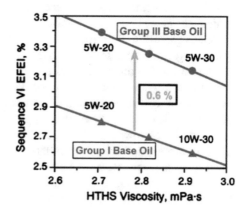

benefit of up to 2 % in the driveline of a vehicle. This results in on overall benefit of up to 5.4 % in a vehicle. In industrial transmissions, it is possible to achieve a 10 % reduction in energy consumption by replacing mineral oil with equi-viscous PAO based oils.

This trend toward improved fuel economy has led to the introduction of lower oil viscosity grades such as 5W30 and 10W30 that are now commonplace in the heavy-duty engines. Analogous changes are also apparent for passenger car engine lubrications where there is a move toward 10W30 and 5W30. Lower oil consumption reduces the soluble organic fraction in the exhaust but lowers the amount of fresh oil and additives to be added to the crankcase. The demands on automotive lubricants have never been greater in terms of their technical requirements.

The base oils of the future will need to possess the following characteristics: thermal oxidation stability, low evaporation, higher natural viscosity index, lower aromatic content, low or no sulfur, low viscosity, and a lower proportion of environmentally damaging materials (metals, sulfur, phosphors and halogens). Low-viscosity synthetic motor oil is thus expected to dominate in vehicles in the coming decades.

3.4 Emissions Reduction Aspect

Reducing the exhaust and CO_2 emissions of cars and trucks constitutes a major challenge for automotive industry. In the current industry climate emissions reduction has been the most significant driver of lubricant quality and the concept of lifetime reduction of emissions is of fundamental importance. Enormous steps have been made in, for example, PM reduction in the last decade or so (Fig. 3.5).

Diesel soot was regarded in the past as carcinogenic due to the adsorption of polycyclic aromatic hydrocarbons (PAH, including nitro and n-PAH). It was concluded that the care of the particle is the main reason. Diesel soot aggregate particles are between 90 and 130 nm in size and aggregated from about 50 primary particles with geometrical diameters between 15 and 30 nm. Spherical primary particulates alone do not occur. Adsorbed on the soot particles are soluble organic as well as gaseous hydrocarbons and sulfates of smaller size.

The share of road transport with regard to PM_{10} emissions, will also be smaller still in the future. Besides the particulate mass the size of the particulate is a further important parameter to consider in terms of health effects. Particles below 2.5 µm can penetrate into the terminal bronchi and below 1 µm into alveolar regions. Heavy-duty diesel engines need low emission lubricants if they are to contribute to the lowering of NO_x and PM emissions. The key elements used in lubricant formulation are sulfur, sulfated ash and phosphorus. These elements will cause compliance problems as emission limits fall. Fuel sulfur level is a major factor for future lubricants because as fuel sulfur levels drop, the contribution of sulfur originating from the lubricant becomes greater. As an example, with a fuel of 10 ppm S, and with a typical mineral-based engine oil with a sulfur level of 0.5 %, an engine with an oil consumption of 0.1 % of the fuel consumption—all numbers quite plausible in the market place today—the equivalent lubricant contribution to sulfur in the exhaust gas in 5 ppm, which is quite comparable to the fuel contribution. Typical sulfur contents (0.1–0.5 %wt) of commercial European car lubricants are shown in Fig. 3.6.

The recently proposed GF-4 specification proposes a lubricant sulfur limit of 0.5 %wt. By avoiding the use of group I base oils, sulfur levels can be reduced to

Fig. 3.5 Typical truck particulate emissions

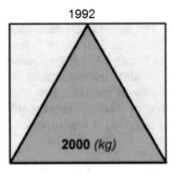

1992

2000 (kg)

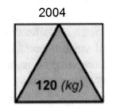

2004

120 (kg)

Fig. 3.6 Composition of current commercial lubricants versus the assumed future limits

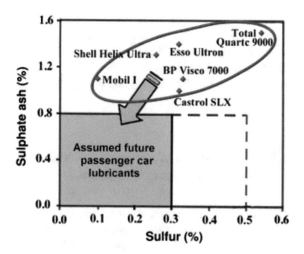

Table 3.3 Sources of SAPS (sulphated ash, phosphorous and sulphur) in lubricants (additives). *Source* Miller et al. (2002)

Component	SAPS contribution
Dispersant	–
Detergents	Ash and sulphur
Anti-oxidants	Sulphur
Friction modifiers	Sulphur
Anti-wear	Phosphorus and sulphur
Diluent oil	Sulphur
Viscosity modifier	–
Corrosion inhibitor	Sulphur

0.3–0.4 % without further changes to formulations, the only consequence being a substantial increase in cost. Lube S (sulphur) and fuel S emissions are the same order of magnitude when using 3 ppm S fuel with lube ranging in S concentration from 0.15 to 0.34 wt% (Table 3.3).

Phosphorus is indicated to impact the activity of gasoline catalytic converters, mainly due to the formation of glassy deposits. Typical phosphorus content of European lubricants is 0.1 %wt. A future phosphorus limit of 0.05 %wt is suggested in the draft ILSAC GF-4 specification. The sulfated ash sources in a lubricant are the metallic detergent zinc di-alkyl di-thiophosphate (ZnDTP). Diesel particulate filters are impacted by sulfated ash built-up. This can lead to a build-up of inert material in after treatment installations, causing a rise in back pressure which upsets programmed air–fuel mixture levels. In extreme cases, accumulation of ash could result in the cracking of ceramic substrates. The typical ash content of European car lubricants is 1.0–1.5 %wt (Fig. 2.4), and 1.0–1.9 % for heavy-duty diesel. Current European engines emit approximately 1.5–3.0 mg/kWh oil ash (corresponding to 8–15 % of total PM for Euro 4/5), and the proclaimed target is a maximum value of 0.5 mg/kWh (corresponding to 2.5 % of total PM for Euro 4/5). Figure 3.3 shows the composition of current commercial lubricants in terms of sulfated ash and sulfur content in comparison to the assumed future limits. The future proposed limit for sulfur and ash is 0.2 wt S (passenger cars) and 0.2– 0.3 %wt S (heavy-duty diesel) and 0.5–0.8 %wt ash (passenger cars) and 1.0 %wt (heavy-duty diesel).

3.5 Contribution of Lubricant Properties to Diesel Exhaust Emissions

Tests of the contribution of lubricant properties to diesel exhaust emissions were performed on a three-cylinder direct injection (DI) engine (THDM 33/T-TD.3.152 Perkins) of rated power 40.5 kW at 2250 rpm and swept volume 2.5 dm^3, turbocharged with intercooler. For those investigated were used two mineral-based motor oils (grade 15W40), and full synthetic (PAO) base oil (grade 5W40).The aim was to show whether the use of synthetic oils rather than mineral ones results in reduction of emissions. Low-sulfur (0.030 %) diesel fuel was used for testing purposes. The resulting physico-chemical characteristics of the oils tested are shown in Table 3.4. The temperature of tested engine oil was 75 °C.

The specific emissions of NO_x and PM (g/kWh) for mineral and synthetic engine oils are shown in Fig. 4.5. It would be seen that NO_x and PM emissions are lower with synthetic oil than with mineral oils. Test results showed that mineral oils produce by 8 % higher NO_x emission than the synthetic oil (Fig. 3.7) This is possibly a consequence of the different additives and aromatic contents of the oils, because there is good correlation between aromatic content and density. At the same time mineral oil has a higher sulfur content (0.8 %) than synthetic oil 0.6 %).

NO_x emission is affected by aromatic content. An increase in aromatic content has been shown to increase NOx and PAH emissions as a consequence of

Lubricating oil	Minaral		Synthetic
	I	II	III
Grade	15W40		5W40
Density (kg/m^3, 15 °C)	880	860	845
Flash temperature	228	226	230
Kinematic viscosity (mm^2/s), 100 °C	15.06	14.6	12.2
Viscpsity index	135	141	160
Noack	10.3	10.1	8.6

Table 3.4 Physico-chemical characteristics of mineral and synthetic base oils. *Source* Plumley (2005)

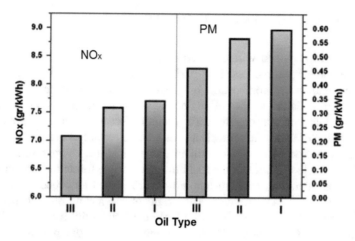

Fig. 3.7 Specific emissions of the mineral and synthetic engine oils

increasing the flame temperature during combustion. On the other hand, decreasing the aromatic and increasing the saturate level provide very thermally stable base oils.

As for particulate matter (PM), it can be noted from Fig. 3.7 that particulate emissions are lower with synthetic oil than with mineral oils. The specific particulate emissions of synthetic oil were 19–24 % smaller than those of the mineral oils. These results are in good compliance with results recently presented in the lubricants literature. Accordingly, switching an older engine from mineral to synthetic lube oil could bring significant benefits in terms of lower emission and fuel economy. Tests carried out by BP in Germany have shown that PAO synthetic oil in some cases reduces particulate emissions by nearly a third and visible smoke by 10 %.

Some investigations demonstrated decreases of between 5 and 22 % in particulate emissions when using synthetic oil as compared to mineral oil. The baseline engine (lubricant) oil was mineral 15W40 oil and the candidate was synthetic 5W30 oil. For operating conditions close to maximum torque, particulate emissions were reduced by 22 %.

It was also shown that the composition of particulate as a function of engine operating conditions. At low speed and load (torque) a high percentage of the oil is found in the particulate. This is because combustion temperature is relatively cool for the oil to burn. As load increases, the oil is less evident. At the high load condition, the predominant change is the non-soluble portion. This means, that under all conditions, the fuel-derived portion of the SOF changed to some degree with the lubricant parameters. In all cases, the synthetic oil reduced the average particulate emissions.

3.6 Lubricant Additives on Particulate Emissions

It is well documented that internal combustion engines produce soot particles during the combustion process. Soot particles provide significant surface area for interaction with other combustion by-products, including organic compounds, ash, and metals. Earlier studies of engine emissions often focused on the relative amount of elemental and organic carbon emitted because both are derived from the combustion of hydrocarbon fuels and oils. More recently, studies have focused on the contribution of lubrication oil to the particle-formation process and, in particular, the role of lubrication oil additives in particle formation. Under normal diesel engine operation, trace metals are vaporized and adsorbed or condensed onto the surface of soot particles. The origin of the metals may be from trace metals in the fuel or from the metallic fuel additives in the fuel used for diesel particulate filter regeneration. However, more typically, the metals in exhaust particles originate from lubrication oil that is spread onto the cylinder walls by the piston rings or that enters the combustion chamber via reverse blow-by of the rings. This is evidenced by data showing measurable levels of metallic lube oil additives in bulk diesel particulate matter (DPM) samples. In diesel engines, the combustion of lubrication oil contributes to particle formation by increasing the amount of semi-volatile hydrocarbon species available for nucleation upon exiting the tailpipe. In addition, the (metallic) ash residues combine with soot particles, and in some cases where the metal-to-carbon ratio is high, metal vapors self-nucleate inside of the engine to form a population of metal-rich nano-particles (Miller et al. 2002).

3.7 Conclusion

This Chapter dwelled on the contribution of engine lubricant oil to diesel engine particulate emissions. It was made evident that mineral engine oil produces higher PM and NO_x emissions than synthetic oil. The NO_x emissions of synthetic engine oil are 8 % lower than those of mineral oil. Particulate emissions of synthetic oil are 19–24 % lower than those of mineral oils. Thus implying that more environmental savings can be achieved by adopting the use of synthetic base oils in the manufacture of lubricants.

Revision Questions

1. *How does lubricant enable fuel economy?*
2. *State two main diesel engine emission forms.*
3. *Why is Group III based oil is preferred in diesel engine oil formulation?*
4. *What are the main advantages of lube extended drainage interval?*
5. *Explain how does more lube emission result from low speed and low load, compared to less at high speed and high load.*

References

ATIEL (2006) Impact of emissions legislation on engine lubricants, emissions, ATIEL, Mar 2006

Madanhire I, Mugwindiri K (2013) Cleaner production in downstream lubricants industry. Lambert Academic Publishing, Saarbrucken, Germany

Miller AL, Habjan MC, Stipe CB, Ahlstrand (2002) Role of lubrication oil in particulate emissions from a hydrogen-powered internal combustion engine. Mechanical Engineering Department, Seattle University, Seattle, Washington 98122

Plumley M (2005) Lubricant formulation and consumption effects on diesel exhaust ash emissions: measurements and sample analyses from a HD diesel engine, 11th Diesel engine emissions reduction (DEER) conference, Chicago, USA, 25 Aug 2005

Chapter 4
Green Lubricant Design and Practice Concept

Abstract Lubricants have an environmental is impact in many ways, but they have a particularly important contribution to make in relation to energy conservation, minimization of waste and development of durable products. Truly green lubricants are those that optimize energy efficiency and minimize wear in the machinery which they lubricate and which have maximized service lifetimes in order to reduce the amount of lubricant required. Increasing importance of these criteria in lubricant selection and design is expected to lead to more widespread use of high performance synthetic base fluids and effective additives. Biodegradability of lubricants from biomass technology may be an area to pursue in terms of research according to this chapter.

4.1 Introduction

There are various sources of air, water and soil pollution but one of the major sources of these types of pollution is lubricants from petroleum based oils due to their poor biodegradability and high toxicity. There is growing concern about the environmental impacts of lubricants. These petroleum based lubricants are poorly biodegradable and have high toxic effect. Spill of these types of lubricants from the working area to the surrounding water or in soil causes very ill effect to the ecological system, which is unacceptable due to rise in environmental concerns. Furthermore these petroleum based lubricants have harmful effects on the operator, lubricants causes various types of skin diseases. Worldwide consumption of these petroleum based lubricants is more than 41 million tons out of which more than 40 % of these lubricants are directly lost in environment through spills and accidents from working machines, which cause very serious damage to the ecosystem. Thus a step towards finding an alternative source of petroleum based lubricants is to be taken for controlling the environmental pollution. This can be achieved by utilizing the available biomass for producing the alternative lubricants (Jain and Suhane 2013).

© Springer International Publishing Switzerland 2016

I. Madanhire and C. Mbohwa, *Mitigating Environmental Impact of Petroleum Lubricants*, DOI 10.1007/978-3-319-31358-0_4

Table 4.1 Oil contents of widely available non edible vegetable oils

Non edible vegetable oils	Oil content by volume (%)
Jatropha	25–35
Karanja	20–25
Castor	37.2–60.6
Mahua	35–40
Neem	20–30

Biomass constitutes the alternative resources available which can be utilized for producing lubricants. Non edible vegetable oil producing plants are the best alternative resource for producing these types of lubricants. These plants are mainly grown for seeds. Oil is extracted from various parts of the plant like leaf, stem, roots but mainly oil is extracted from seeds of the plant. On an average seeds contains about 40–60 % of oil which can be used as a base fluid for making lubricant formulations and the specific yield are given in Table 4.1 (Jain and Suhane 2013). Some of the widely available non edible vegetable oils producing plants are Jatropha (*Jatropha curcas*), Karanja (*Pongamia pinnata*), Mahua (*Madhuca indica*), Neem (*Azadirachta indica*), Simarouba (*Simarouba glauca*), Wild apricot (*Prunus armeniaca*), Castor (*Ricinus Communis* L) etc. Non edible vegetable oil can also be used as an additive or as a blend with petroleum based lubricants. Lubricants from these types of alternative resources are generally known as bio-lubricants.

4.2 Vegetable Bio-lubricants

Bio-lubricants is a revolutionary step in the world of lubricant. Bio-lubricants are described in many ways like eco-friendly lubricants, green lubricants, biodegradable lubricants, recyclable, nontoxic and re-usable etc. Bio-lubricants compared to petroleum based lubricants have very low or no pollution causing effect on environment. Bio-lubricants also have zero toxicity, so no skin diseases to the operator in case of leak or accident. Bio-lubricants are highly biodegradable compared to petroleum based lubricants. Furthermore bio-lubricants possess almost similar features like viscosity, lubricity, viscosity index etc. Which makes bio-lubricants a best alternative of petroleum based lubricants.

Vegetable oil as a lubricant is preferred not only because they are renewable raw materials but also because they are biodegradable and non-toxic. They also acquire most of the properties required for lubricants such as high index viscosity, low volatility and good lubricity and are also good solvents for fluid additives. However, vegetable oils have poor oxidative and thermal stability, which is due to the presence of unsaturation. This unsaturation restricts their use as a good lubricant. Several attempts have been made to improve their oxidative stability such as transesterification of trime-thylopropane and rapeseed oil methyl ester; selective hydrogenation of polyunsaturated C=C bonds of fatty acid chains and conversion of C=C bonds to oxirane ring via epoxidation. Among these, epoxidation received special attraction because it opened up a wide range of feasible reactions that can be carried out under moderate reaction conditions due to the high reactivity and functionality of the oxirane ring. For instance, the epoxides can react with different nucleophiles to produce mono-alcohols, diols, alkoxyalcohols, hydroxyesters, N-hydroxyalkylamides, mercaptoalcohols, aminoalcohols, hydroxynitriles, etc.

As given in Fig. 4.1, the life cycle of lubricant polymers based on vegetable oils, according to which the biomass from plant-derived resources is extracted in order to yield the vegetable oil. Subsequently, the oil is submitted to chemical modification with the aim of enhancing its reactivity towards a given type of polymerization approach. The polymers are then made available to the consumers, and once used, they become waste, which after degradation and assimilation is reused as biomass and the cycle starts again (Samarth et al. 2015).

It was established that epoxidized soyabean oil as a potential source for high temperature lubricants applications. Also product obtained by ring-opening reaction of epoxidized fatty acid esters followed by esterification of the resulting hydroxyl group shows good performance for lower temperature lubricant applications. A process for the production of biodegradable lubricant-based stocks from epoxidized vegetable oil with a lower pour point via cationic ion-exchange resins as catalysts was also developed. This involves two steps, first, ring-opening reactions by alcoholysis followed by esterification of the resultant hydroxy group in the first step.

The triglyceride structure of vegetable oils provides qualities desirable in a lubricant. Long, polar fatty acid chains provide high strength lubricant films. The strong intermolecular interactions are also resilient to changes in temperature providing a more stable viscosity, or high viscosity coefficient. The entire base oil is also a potential source of fatty acids. The triglyceride structure is also the

Fig. 4.1 Life cycle of polymeric material based on vegetable oils (adapted from Samarth et al. 2015)

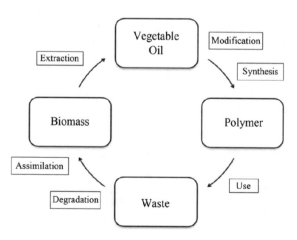

basis for the inherent disabilities of vegetable oils as lubricants. Unsaturated double bonds in the fatty acids are active sites for many reactions, including oxidation, lowering the oxidation stability of vegetable oils. Another concern is the susceptibility of the triglyceride ester to hydrolysis. The similarities in all vegetable oil structures mean that only a narrow range of viscosities are available for their potential use as lubricants. Finally, the strong intermolecular interactions whilst providing a durable lubricant film also result in poor low-temperature properties.

Two new synthetic routes to polyol-derived intermediates, were proposed as an alternative way for the production of lubricants from vegetable oils obtained from temperate climate crops such as soybean and sunflower seeds. The oils are first epoxidized, and then the oxirane ring is opened with either acetic acid or a low chain aliphatic alcohol (methanol or ethanol). These routes are promising in that they demand short reaction times and low reaction temperatures, while 99 % conversion is achieved, with high selectivities. Considering only the viscosity of the products, they are all viable lubricant base stocks in which the intermediates can be further esterified to obtain low-temperature lubricants. The development of soybean oil-based lubricants, the preparation of synthetic lubricant base stocks from epoxidized soybean oil and Guerbet alcohols was also reported. The pour point, viscosity, viscosity index, and oxidative stability in micro oxidation test were evaluated to investigate the effect of structural variation in the oil molecules (Samarth et al. 2015).

Biodegradability is one of the most important property of lubricant; a good lubricant should be highly biodegradable when released in environment. Biodegradability of lubricant is defined as "susceptibleness of a lubricant to degrade under the effect of biological organism like fungi without causing any diverse effect on environment." Petroleum based lubricants are poorly biodegradable while the lubricants from alternative resources are highly biodegradable and do not cause any adverse effect on the environment as given in Table 4.2.

From the Table 4.2, lubricants from alternative resources are having much higher bio-degradability than lubricants from petroleum based lubricants. Lubricants from alternative resources are having biodegradability in the range of 90–100 %, which means that these lubricants quickly degrade in the effect of biological organism. Thus the adverse effects of lubricants caused to the environment are reduced at a greater extent, which is the demand of today's world. Also they found there use in various industrial applications.

Table 4.2 The biodegradability of petroleum based lubricants and lubricants from alternative resources

S.No.	Lubricant	Biodegradability (%)
1	Petroleum based lubricants	20–30
2	Lubricants from alternative resources	90–100

Use of modified vegetable oil is a convenient way toward the goal of green chemistry, and is strongly recommended to use in polymer area. The examples cited above are impressive and provide a good insight into the field of utilization of vegetable oil as polymeric material. In order to achieve further development in this field, improved method and modification, which give rise to better properties and constitute a minimal hazard, should be used instead of the petrochemical based material.

4.3 Environmental Pollution Control

Controlling environmental pollution caused by petroleum based lubricants is of prime importance, as these lubricants having adverse effect on the ecological system and on all the living beings. Comparison of petroleum based lubricants with lubricants from alternative resources is shown in Table 4.3. It is clear that petroleum based lubricants are the main cause of Environmental pollution, while the lubricants from alternative resources are best to reduce the environmental pollution and also to reduce the harmful effects on the human being and other living beings.

Alternative resources like non edible vegetable oils can be produced domestically while the petroleum oil cannot be produced domestically and it also on the verge of depletion so as its price is also increasing. Petroleum based lubricants are dangerous to handle and store in case of spill, leakage etc. due to their poor biodegradability and high toxicity while this is not the case with the lubricants from alternative lubricants. For diesel engines high temperatures cannot be matched by current bio-lubricants. Thus synthetic oils are the only way to achieve extended drain interval.

Table 4.3 Comparison of petroleum based lubricants with lubricants from alternative resources

Environmental pollution control			
No.	Aspect description	Petroleum based lubricants	Lubricants from alternative resources
1	Biodegradable	No	Yes
2	Toxic	Yes	No
3	Harmful by-products of emission	Yes	No
4	Cause of global warming	Yes	No
5	Highly toxic to humans and animals	Yes	No
6	Pollutes environment	Yes	No
7	Economic gain to Indian farmers	No	Yes
8	Dangerous to handle and store	Yes	No
9	Domestic production	No	Yes
10	Alternative resources	No	Yes

4.4 Lubricants for High Temperature Diesel Engines

A case in point is the fact that mineral-oil based lubricants are being successfully used in current high temperature diesel engines. The maximum ring zone temperature has been measured by temperature-sensitive plugs to be about 350 °C. At this temperature, no mineral oil-based lubricant should survive, yet mineral-oil based lubricants are being used satisfactorily in such conditions. This is because of the short residence time for which the lubricant is subjected to such temperatures. The residence time of the lubricant in the ring zone has been measured recently to be about three minutes. So, if a lubricant can still protect for three minutes, the lubricant is adequate. We can define this conceptually as a kinetically retarded lubricant. In order to design a lubricant that can survive the longest time at the maximum design temperature, base fluid selection and additive combinations are the critical issues.

4.5 Synthetic Lubricants and Long Drainage Intervals

Synthetic PAO (poly alpha olefin) hydrocarbons are made in a process that results in the desired types of hydrocarbon molecules, isoparaffins. Synthetic lubricants can be used in higher temperatures than conventional oils without breakdown. This resistance to breakdown allows synthetic oils to be utilized longer than conventional oils in addition to systems being cleaner and lasting longer. It is thus possible through a chemical reaction to produce the best possible lubricating oil, which entirely lacks the unwanted components. This is the most commonly used synthetic base oil in modern engine lubricants.

At present, used motor oil is the largest source of oil pollution in waterways. The first thing we can do is to not create so much used oil to begin with. Maybe it is time to take advantage of the benefits that synthetic lubricants provide. Besides the financial benefits we attain from synthetic lubricants, we can protect our precious environment as well. The future of mineral-based lubricating oil is limited, because the natural supplies of petroleum are both finite and non-renewable.

Experts estimate the total recoverable light to medium petroleum reserves at 1.6 trillion barrels, of which a third has been used. Thus, synthetic-based oils will probably be increasingly important as natural reserves dwindle. This is true not only for lubricating oil but also for the other products that result from petroleum refining. The result is that you will generate far less waste oil (and empty oil containers), resulting in a considerable reduction in environmental contamination and damage.

Fig. 4.2 Thermal stability of additives

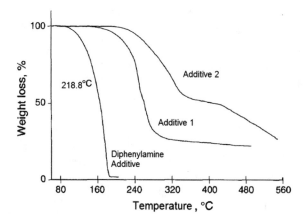

4.6 Additives to Match High Temperatures

The availability of thermally stable additives is a major problem in developing a super-stable lubricant. The requirements on the additives include good thermal and oxidative stability, solubility, and compatibility with other additives. These additives must be effective at high temperatures as well as in the sump. In addition, the additives should not contribute significantly to deposits in the top ring reversal (TRR) zone. The additives considered included antioxidants, dispersants, surface deactivators, and anti-wear additives.

A number of additives whose additive chemistries had been used in applications other than lubricants, due either to solubility or to incompatibility. These were screened for their thermal stability range resulting in some typical antioxidant thermo-grams shown in Fig. 4.2. If the additive meets the thermal stability requirement, i.e., it will not completely disappear within the temperature and time target, it will be tested for its functionality effectiveness in a multi-component additive blend designed for component evaluation. Some additives found very effective at the targeted temperatures were not normally used with lubricants, yet they have much higher thermal stability. In one particular case, the lack of solubility of the additive in hydrocarbons was the reason.

4.7 Lube Deposit Formation

The deposit-forming tendencies of the lubricants in high temperature diesel application have been a major concern. It has been shown that top land carbon deposits can result in loss of oil control. Loss of oil control can also result in increased engine emissions. To evaluate deposit-forming tendencies of lubricants, a 'two-peak' PDSC method was developed to evaluate lubricant deposits at high

Fig. 4.3 Typical PDSC
2-peak deposit method
thermo-gram

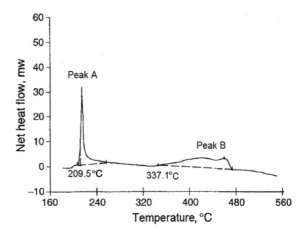

temperatures. Basically, a thin film of lubricant is oxidized in oxygen at 690 kPa over a temperature ranging from room temperature to 550 °C.

At high temperatures, the nature of the metal surface is a critical parameter. The shape and size of the reactor that is to allow rapid oxidation without volatility and evaporative losses. The result is a thermo-gram like the one shown in Fig. 4.3, where the first peak is the primary oxidation peak and the second one is related to the deposit-forming tendencies of the fluid.

The key to this bench test is in controlling the competing thermo-oxidation reactions and the evaporative and volatility losses. If too much material is lost as a result of volatility, there is an insufficient amount left to form a deposit. If insufficient oxygen is allowed into the system, then the oxygen diffusion rate becomes the limiting factor. Catalytic activity of the iron surfaces also plays a role and tends to accelerate the oxidation reactions. This two-peak method has been correlated with engine test data obtained from four companies.

4.8 Hot Metal Surface Effects

The presence of metal surfaces in lubricants in thermal oxidative conditions has been shown to accelerate the oxidation degradation of lubricants. To evaluate the effect of metal surfaces at elevated temperatures and the effectiveness of high-temperature surface deactivators, an isothermal PDSC procedure at 200 °C, with alternate pans, was run. At temperatures above 200 °C, the hot metal surface exerts an overwhelming effect on lubricant degradation. Once an additive was found to be effective, it was then tested in various combinations of additives for additive compatibility. Many of the metal deactivators are effective at temperatures below 180 °C, but are ineffective at higher temperatures.

4.9 Environmentally Considerate Lubricants (ECL)

Avoiding lubricants entering the environment in the first place is the ideal scenario, but unfortunately, even in the most carefully planned and well managed operation, some amount of leakage is inevitable. For example, hydraulic systems on mobile mining equipment usually operate at high pressures with exposed hoses, cylinders and seals. As such, there is a high risk of hoses breaking and hydraulic fluid entering the environment. Even a leak of one drop per second from a machine can give a cumulative fluid loss over a month of about 200 L. To ensure reduced environmental impact, companies can select from a range of lubricants which are environmentally considerate and have the ability to biodegrade into harmless, natural substances.

Biodegradability is the ability of a lubricant to be broken down by microorganisms, such as bacteria and fungi, to simpler compounds. This is the alteration in the chemical structure of a substance, brought about by biological action, resulting in the loss of specific properties of that substance. It does not mean that complete biodegradation has occurred, as the pollutant might still be present in different chemical forms. Ultimate aerobic biodegradation is the level achieved when the test compound is totally utilized by micro-organisms, resulting in the production of harmless substances such as carbon dioxide, water, mineral salts and new microbial cells. A substance which is "readily" biodegradable has passed certain specified screening tests for ultimate biodegradability. A lubricant for use in an environmentally sensitive application should combine a number of important properties.

These include high biodegradability, which means it is rapidly removed from the environment by natural processes in the event of a leak or spill, and low eco-toxicity. Eco-toxicity is the potential for a material to produce adverse effects in animals and plants. The eco-toxicity implications of a lubricant are more complex and diverse than the biodegradability implications, and are usually dependent on the additive composition. In addition to coal mining, applications for ECLs include machinery operating in sectors such as forestry, agriculture, construction, railways, earth-moving, marine and water treatment. In all these sectors, the consequences of a major spill could be serious, and inevitable routine leakages accumulate over time, with the potential to cause significant environmental damage. Forestry is an example of an industrial sector which has recognized the potential environmental consequences of its activities (Madanhire and Mugwindiri 2013).

Forestry companies consume vast quantities of oils and greases, many of which could affect the environment. Environmental concerns have led to environmentally considerate lubricants being made compulsory in Swedish and German forests. It is now clear that ECLs can play a major role in environmental protection programs on a global basis.

4.10 Recycling and Reclamation of Lubricants

Improvement in filtration technologies and processes has now made recycling a viable option (with rising price of base stock and crude oil). Typically various filtration systems remove particulates, additives and oxidation products and recover the base oil. The oil may get refined during the process. This base oil is then treated much the same as virgin base oil however there is considerable reluctance to use recycled oils as they are generally considered inferior.

Base stock fractionally vacuum distilled from used lubricants has superior properties to all natural oils, but cost effectiveness depends on many factors. Used lubricant may also be used as refinery feedstock to become part of crude oil. Again there is considerable reluctance to this use as the additives, soot and wear metals will seriously poison/deactivate the critical catalysts in the process.

Cost prohibits carrying out both filtration (soot, additives removal) and re-refining (distilling, isomerization, hydrocrack, etc.) however the primary hindrance to recycling still remains the collection of fluids as refineries need continuous supply in amounts measured in cisterns, rail tanks.

4.11 Extended Condition-Based Drainage Interval

The decision depends upon several factors, including machine criticality and failure history, tank size, accessibility for lubrication maintenance, desire for other information provided by oil analysis, including contaminant levels and wear debris information, etc. Extending the interval too far places the machine at-risk for wear and failure due to underperforming lubricant and/or excessive contamination levels. Factors such as operating temperature, presence of water contamination, aeration levels, contaminant ingestion rate and wear generation rate, along with propensity for risk and planning and scheduling windows all influence the decision.

Where oil analysis is employed, one must decide upon the test slate, which should reflect the condemning factors thought to represent a decline in the lubricant's performance properties, the caution and condemning limits for the selected test slate, the sampling and analysis interval, the sampling method and system, including required machine hardware modifications, training and certification for the staff to enable effective sampling and diagnostics and the oil analysis information management system.

4.12 Leakage Management

Leaky machines can cause injuries, fires, improper/slowed operation, quality defects and environmental damage, not to mention high labor and material costs. Machines leak due to improper design, operation or maintenance. Ideally, leakage should be managed by identifying its source and cause and by taking corrective actions to eliminate it. Elimination of leakage is not always feasible and, therefore, the leaks must be effectively dealt with to limit the damage they cause. Containment and guttering systems for leaked oil can be expensive to install and maintain, and they may be only marginally effective. Where possible, leakage elimination is preferred.

4.13 Future of Green Lubricants

Lubricants from alternative resources control the environmental pollution by preserving ground water and soil, reducing petroleum oil consumption, cost of disposal is reduced, natural recycling of hazardous waste, high flash and fire point reduce risk of fire in case of accident and no skin problem to the operator. Biotechnology has a vital role to play in reducing the environmental pollution thus utilizing the alternative resources. The adverse effects of petroleum based lubricants on environment and all living beings can be reduced at a greater extent by promoting the bio-lubricants. There is a need to take right step in the field of development of lubricants from alternative resources to make it more cheap and easily available, it can be achieved by focused research through appropriate government polices and regulation for the use of lubricants from alternative resources.

4.14 Conclusion

Green lubricant concept is a variety of lube related efforts and activities to mitigate impact of lubricants on the environment. These include bio-lubricant base oils, biodegradability, synthetics for high temperature diesel, oil extended drain intervals, environmental considerate lubricants (ECLs), oil recycling and reclamation, and leakage management. The latest compromise was a synthetic bio-lubricant the polyol polyalpha olefin, which still has its own shortcoming on hydrolysis attack. As a result, more research is required to ensure that bio-lubricants are modified to be suitable for use in a range of industrial application with high temperature with minimum oxidation. Currently the synthetics for these severe applications, although they enable extended drainage interval of 100,000 km, are not readily bio-degradable.

Revision Questions

1. *How does mineral base oil pollute the environment?*
2. *Which common vegetative plant is used to produce a bio-lubricant?*
3. *What bio-degradability level is required for a bio-lubricant?*

4. *Currently vegetable bio-lubricants are not suitable for use in high temperature diesel engine. What type of base oil is used and why?*
5. *List four aspects which constitute the process of stabilizing bio-lubricants.*

References

Hsu SM, Perez JM, Ku CS (2006) Advanced lubricants for heat engines. National Institute of Standards and Technology, Gaithersburg, Maryland, USA

Jain AK, Suhane A (2013) Mini review biotechnology: a way to control environmental pollution by alternative lubricants. Res Biotechnol 4(3):38–42, ISSN: 2229-791X, www.researchin biotechnology.com

Madanhire I, Mugwindiri K (2013) Cleaner production in downstream lubricants industry. Lambert Academic Publishing, Saarbrucken, Germany

Maleville X, Faure D, Legros A, Hipeaux JC (1996) Oxidation of mineral base oils of petroleum origin: chemical composition, thickening, and composition of degradation products, Lubrication science 9(1), November 1996

Reeves CJ, Menezes PL (2016) Advancements in eco-friendly lubricants for tribological applications: past, present, and future. Ecotribology research developments—materials forming, machining and tribology. vol 2, Springer International Publishing, Switzerland, pp 41–61 http://www.springer.com/978-3-319-24005-3. doi:10.1007/978-3-319-24007-7_2

Samarth NB, Prakash A, Mahanwar PA (2015) Modified vegetable oil based additives as a future polymeric material—review, Open journal of organic polymer materials, vol 5, Scientific Research Publishing, pp 1–22 http://www.scirp.org/journal/ojopm

Chapter 5
Synthetic Lubricants and the Environment

Abstract The benefits of synthetic lubricant base stocks are derived not only from the basic molecular structures but from the absence of harmful molecular species often unavoidably present in conventional mineral oils in small, but significant, concentrations. There are very many compounds in crude oil, and while many, or most, of the harmful ones are removed or upgraded by refining, depending on the methods used, a significant number will inevitably remain in lubricating oil stocks, whether solvent- or hydro treated. Thus, conventional oils comprise a wide variety of molecular species, many of which are not well characterized. Synthesized hydrocarbons are now used for a wide range of industrial and automotive applications and are, by far, the segment with the greatest growth rate in the synthetic lubricant field.

5.1 Introduction

Strict new environmental regulations have directed industry to reduce the demand on limited natural resources, reduce industrial waste and minimize hazardous emissions. On the one side, there are more systems that require lubrication and, on the other side, consumption must be reduced. Synthetic lubricants were developed and used for applications in which mineral base oil products were inadequate such as at extremely high and low temperatures, under extreme wear conditions, or where special characteristics such as long life, improved equipment efficiency, or non-flammability, were needed. However, the use of synthetic lubricants is not restricted to special applications, but is also useful when they can provide cost efficiency in areas such as machine reliability, oil life, energy consumption, biodegradability and safety. Among many successful applications, typical cases where synthetic fluids have clear advantages over conventional lubricants include significant lengthening of oil-change intervals, fewer machine shutdowns and environmental protection (Murphy et al. 2002).

© Springer International Publishing Switzerland 2016   59
I. Madanhire and C. Mbohwa, *Mitigating Environmental Impact*
of Petroleum Lubricants, DOI 10.1007/978-3-319-31358-0_5

Used oil by its nature, generated by its use as a lubricant in automotive and industrial operations, constitutes a hazardous waste stream. Source reduction, which is preventing or reducing the amount of waste generated, is the best option for managing waste, according to the waste management hierarchy. Engine oil performs under harsh conditions inside an engine with its combination of heat and high pressure, combustion activities and generation of chemical residues. In this harsh operating environment, the oil gets dirty, additives and other chemicals break down, and the oil requires regular changing.

The amount of used oil generated by the use of lubricating oil in motor vehicles can be reduced by increasing the number of kilometers driven between oil changes. Today's synthetic lubricants are marketed as providing the consumer with the convenience of extended oil change intervals, in addition to offering better engine performance. To address the potential environmental and human health impacts associated with the use of these synthetic lubricants, this chapter reviews on their chemical, physical and toxic properties, along with trends in their use.

5.2 Synthetic Versus Mineral Lubricants

Mineral base oils are manufactured by the distillation of crude oils, followed by further refining of the distillates via separation or other conversion processes (e.g., hydro cracking, hydrogen reforming, and wax isomerization). These oils are mixtures of paraffins (straight- or branched-chain hydrocarbons), naphthenes (ring forms of paraffins) and aromatics (alkyl benzenes and multi-ring aromatics). Mineral base oils break down in extreme heat and congeal in extreme cold. In addition, mineral oil base stocks contain undesirable impurities such as sulfur, trace metals and carbon residues which can limit the performance capabilities and useful service life of the resulting blended lubricant oils (Wills 2005).

Unlike mineral oil base stocks, which are a complex mixture of naturally-occurring hydrocarbons refined from crude oil, synthetic base stocks are man-made, having controlled molecular structure with predictable properties. As with all present-day fluids lubricants, synthetics are formulated by combining the base stocks with selected additives. Quite often, the selection of a given synthetic lubricant for a specific application is determined by one or more outstanding properties of that synthetic which are beyond the capability of the conventional petroleum lubricants and/or where the additional cost can be justified.

Synthetic base oils can be substituted for conventional mineral base oils. Most synthetic motor oils are fabricated by polymerizing short chain hydrocarbon molecules called alpha-olefins into longer chain hydrocarbon polymers called polyalpha olefins (PAOs). The degree of variation in molecular size, chain length and branching in synthetically produced fluids is much less than occurs in base stocks extracted from crude oil. While they appear chemically similar to mineral oils refined from crude oil, PAOs do not contain the impurities or waxes inherent in conventional mineral oils. PAOs constitute the most widely used synthetic motor oil in the U.S. and Europe (Denton 2007).

Synthetic lubricant technology allows the products to be designed for particular lubrication applications and, in combination with additives, provides targeted performance. Synthetic lubricants are frequently blended with mineral oil in order to provide desired combinations of properties. Among the advantages for synthetic lubricants over mineral base oils are: low temperature fluidity and thus better cold weather performance; low volatility (i.e., low tendency to evaporate); high-temperature thermal stability (high "viscosity index"); oxidation resistance (of the oil itself); and high natural detergent characteristics (resulting in a cleaner engine with less additive content). The "green" advantages that accompany the use of synthetic oils include improved fuel economy, decreased oil consumption and extended oil change intervals.

5.3 Synthetic Base Oil Classification

The American Petroleum Institute (API) classifies base oils under five categories these are Group I, Group II, Group III, Group IV and Group V. These categories help identify base stocks in finished oil formulations to ensure that engine oil performance demands are met (Denton 2007). API has classified synthetic engine oils made with PAOs as a special class of base stock. The term "synthetic" was originally used to refer to Group IV (PAOs) and V base stocks. With the growth of the PAO market, some base oil manufacturers began manufacturing Group III mineral oils that provide equivalent performance to PAOs, and marketing these as "synthetic" oils. In 1999, a ruling by the National Advertising Division of the Council of Better Business Bureaus broadened the definition of "synthetic lubricants" to include high-performing products made with Group III base stocks shown in Fig. 5.1 for high quality low viscosity oils to achieve fuel economy.

Fig. 5.1 Viscosity grade/ base stocks on fuel economy

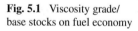
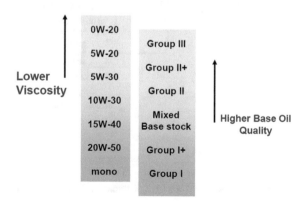

5.4 Demand for Thermal-Oxidative Oils for High-Temperature Diesel Lubricants

Many advanced engine concepts require lubricants that function in a high temperature environment. Thus thermal-oxidative characteristics of some lubricants formulated were considered for improvement in thermal stability, oxidative stability and oxidative volatility for lubricants designed for high temperature applications. Engine lubricants are normally complex mixtures of hydrocarbons. When the hydrocarbons are subjected to high in-service engine temperatures, free radicals are generated via either hydro-peroxide and/or carbon-carbon chain scission (thermal degradation mechanism) as shown by Fig. 5.2. These reactions yield complex intermediates, including ketone, alcohols, and carboxylic acids which subsequently decompose into smaller molecule fragments (low molecular weight products (LMW), compared with the original), or polymerize into larger molecules (high molecular weight products (HMW)). The smaller molecules formed could evaporate or undergo polymerization reactions to form high molecular weight products. It in this regard that new lubricant technology is needed for advanced engine concepts. The proposed low-heat-rejection (LHR) engine has the advantage of low particulate exhaust emissions, high fuel efficiency, and high power density.

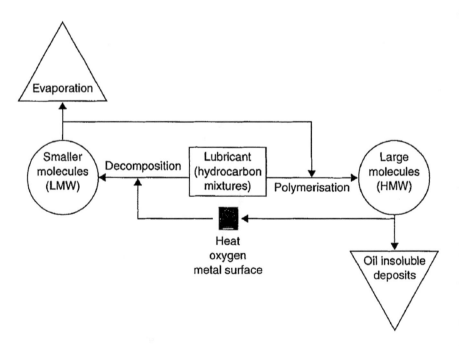

Fig. 5.2 A generalized lubricant degradation diagram (Li et al. 2010)

Volatility and oxidative volatility are other issues greatly affected by the higher temperatures. Lubricants have low vapor pressures under the normal engine operating temperature range of 100–275 °C, but when the temperature is high enough to boil half of the lubricant fraction, physical evaporation will be significant. This critical temperature generally is slightly higher than 300 °C for a mineral oil and can be much higher for a synthetic lubricant base according to tests done. Thus, higher temperatures accelerate oil consumption greatly. To minimize oil volatility and increase the oil drain interval, new types of base stock are being used. These generally fall into the Group II, III, and IV API base oil categories (Hamblin and Rohrbach 2001).

The demand for lubricants to function at high temperatures is causing a move from mineral-oil-based lubricants to synthetic lubricants. At higher temperatures, as indicated, several phenomena may have increased influence on the performance of a lubricant. These phenomena include: volatility, deposit forming tendency, and hot metal surface catalytic effect. In an engine, the rate of lubricant degradation is controlled by the oxygen diffusion rate, temperature, and metal surfaces acting as oxidation catalysts. When the engine temperature rises, a harsher environment is created for the lubricant. At such temperatures, synthetic lubricants are needed to serve the engine (Li et al. 2010).

5.5 Lubricants Based on Synthesized Fluid

The characteristics of synthetic lubricants are derived from the physical and chemical properties of the base fluids and from the effect of the additives used. The physical and chemical qualities of the base fluids include such things as viscosity-temperature behavior, low-temperature fluidity, volatility, traction, compatibility with paints and sealants, miscibility with petroleum, hydrolytic stability, and the ability to dissolve chemical additives. Additives are often used in synthetic lubricants and they can influence to a greater extent, oxidation stability, load-bearing ability, corrosion protection, and foaming and demulsifying performance. New products that are developed, for example, environmentally adapted lubricants (EALs), have different fundamental properties from conventional mineral-oil based products.

5.6 Hydrocarbons Build-up from Mineral Oil-Based Lubricants

Synthetic products are normally produced by chemical synthesis reactions of very pure, small molecules to produce equally pure synthetic base oils. Frequently, the synthesis of the reaction up to the desired end product includes several steps, each of which necessitates a purification of the intermediate products. It is this

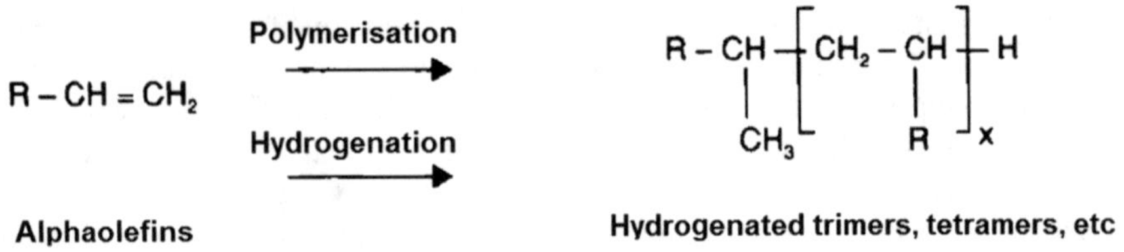

$$R - CH = CH_2 \quad \xrightarrow{\text{Polymerisation}} \quad \xrightarrow{\text{Hydrogenation}} \quad R - CH \left[CH_2 - CH \right]_x H$$

Alphaolefins

$$\underset{CH_3}{\overset{}{\mid}} \qquad \underset{R}{\overset{}{\mid}}$$

Hydrogenated trimers, tetramers, etc

Fig. 5.3 Poly alphaolefin reaction

resulting freedom from undesirable compounds that gives synthetic lubricants their distinctive characteristics. It also contributes to their higher cost, compared to mineral oils. The net cost of a synthetic product is the sum of the costs of the raw materials and the costs of the individual process steps. The synthesis routes of the major synthetic base oil types are described in Fig. 5.3.

Poly alphaolefins (PAOs) are paraffin-like liquid hydrocarbons that represent one structure of a class of fluids known as synthesized hydrocarbons that comprise molecules containing only carbon and hydrogen. The major application of PAOs is as a lubricant base stock. This class of synthetic fluid closely resembles mineral oil in many properties, but with key enhancements. PAOs exhibit a unique combination of high-temperature viscosity retention, excellent shear stability, low volatility, very low pour point, and an excellent response to antioxidant additives, resulting in very high oxidation-resistance potential. To improve thermal and oxidative stability, the remaining double bonds in the molecules must be removed. This is accomplished through hydrogenation, the addition of hydrogen, which saturates the molecule and eliminates double bonds. Intermediate steps include washing to remove catalyst and distillation to achieve the desired viscosity.

5.7 Health Impact on Humans

5.7.1 Toxicity of Unused Lubricating Oils

Lubricating oils are viscous liquids with low vapor pressures and low volatile organic compound (VOC) content. Thus, exposure to hazardous VOCs via inhalation is expected to be minimal. Both mineral-based and synthetic oils have low acute oral and dermal toxicity. The main effects in humans following accidental ingestion of even large quantities of these oils are limited to irritation of the digestive tract, with symptoms of nausea, vomiting and diarrhea. Skin may be mildly or moderately irritated following repeated or prolonged exposure to mineral and synthetic base oils (accidental spills rarely cause problems). Repeated contact can cause defatting of the skin and give rise to signs of irritancy, i.e., redness, inflammation and cracking. These health effects may be attributed to the additive components (e.g., metals and detergents) in lubricating oils.

Unused mineral-based lubricant oil contains very small amounts of polycyclic aromatic hydrocarbons (PAHs). A number of PAHs are classified as "probably carcinogenic to humans" based on animal evidence. Mild hydro treating (i.e., Group II oils) helps reduce the amounts of carcinogenic PAHs but does not necessarily eliminate them. Increasing the temperature and pressure of hydro processing can eliminate carcinogenic compounds. Of the refining steps used in preparing lubricating oil base stocks from petroleum, only effective solvent extraction, severe hydrogenation or exhaustive fuming sulfuric acid treatment appear adequate to eliminate PAHs. Newly synthesized PAOs (Group IV base stocks) do not contain PAHs.

5.7.2 Toxicity of Used Lubricating Oils

In addition to the mixture of hydrocarbons and additives present in the formulated product, used crankcase oil contains contaminants that accumulate during use as an engine lubricant. Sources of contamination include additive breakdown products (e.g., barium and zinc); engine "blow-by" (i.e., material which leaks from the engine combustion chamber into the crankcase oil); burnt oil; and metal particles from engine wear, such as arsenic, lead, nickel and cadmium. Numerous other metals are present in used oils such as aluminum, copper, iron, magnesium, silicon and tin; however, they are generally not given much attention due to their low concentrations and low toxicities (Denton 2007).

Motor oils become "enriched" with PAHs during the operation of an engine. These contaminants are fuel combustion products that are transported into the crankcase and concentrate in lubricating oil. In an early study using a 1981 gasoline-powered vehicle, PAHs were not detected in new lubricating oil; however, concentrations increased rapidly with increased kilometers driven. PAH concentrations (predominantly two- and three-ring compounds) increased until about 6436 km and then leveled off. The more toxic five ring PAHs (benzopyrenes) were only detected at 9332 km, the longest distance driven. At the end of the study the total PAH concentration in the used oil was 10,300 μg/g or about 1 % of the oil. It was concluded that PAHs accumulate in crankcase oil with increasing kilometers driven. In some cases, these studies measured PAH levels in crankcase oil drained at mileage intervals that may be lower than the manufacturer's recommended oil change intervals for today's vehicles and lubricants.

5.7.3 Effect of Extended Drainage Interval on Used Oil

As previously stated, the use of highly refined base oils and synthetic oils provides an opportunity for source reduction due to the extended oil change intervals. Depending on the synthetic oil used, oil change intervals can range from

12,000 km to as high as 40,000 km. Because PAHs have been shown to concentrate in oil during its service life, one of the potential implications of these longer drain intervals is that used synthetic engine oil may contain more PAHs and therefore pose a greater risk to human health. For example, studies indicate there are higher PAH levels in particulate emissions from engines operating with used motor oil compared to fresh lubricant oil. In this case, people might be exposed via inhalation to elevated levels of PAHs adsorbed to particulates formed from the combustion of lubricant oil and fuel in the cylinders. The recycling of used oil with higher PAH levels to produce fuel oil for combustion may similarly result in higher exposures to PAHs.

Additionally, exposure to PAHs (and other contaminants, notably metals) in used engine oils might occur from dermal contact while changing oil as well as from handling recycled oil used as fuel.

5.8 Environmental Impacts

Used oil that is leaked, spilled or improperly discarded may find its into stormwater runoff and eventually enter into and adversely affect the environmental health of receiving water bodies. Studies monitoring contaminants in runoff consistently report relatively low levels ($\leq$5 mg/L) of oil and grease entering into surface waters. It has been reported that petroleum hydrocarbons in urban runoff as well as in aquatic sediment in urban areas are primarily associated with used crankcase oil. However, as reported in 2006, the extent to which used motor oil and oil byproducts are polluting stormwater runoff and the ultimate receiving waters is largely unknown. In the case of the highly refined motor oils and synthetic lubricants, the increased PAH levels accumulating due to extended drain intervals (as discussed above) may result in increased used oil related PAHs entering into stormwater runoff. This may, in turn, result in higher concentrations of PAHs in our nation's rivers, bays, oceans and sediments.It was also established that greenhouse emissions of PAO are almost twice than those of mineral base oil, due to higher quantities of refinery gas burned for heat consumption and, in general, to a more energy consuming production process.

However, in reason of the fact that modern engines requires lubricating oils that can lead to higher performance, reducing frictions and fuel consumption, this can lead to environmental benefits in a life cycle perspective. Synthetic oils offer a longer life time and require less oil changes, leading to a decrease of environmental impacts per distance covered. Hence there is room for improvement in the production of additives and fully formulated lubricants through the deployment of new technologies such as use of nano-particle material (Girotti et al. 2011).

5.9   Advantages of Synthetic Lubricants

PAOs have very good general performance properties with distinct advantages over mineral oils, plus a wide range of available viscosity grades. In addition, PAOs are miscible with mineral oils. PAO-based products have important advantages over hydro processed mineral base oils that have demonstrated some benefits compared with conventional mineral oils. For most industrial applications, the availability of high-viscosity PAOs provides a singular advantage over hydro-processed mineral stocks that are generally limited to the lighter grades. Additional performance advantages of PAOs over these stocks include traction, efficiency, oxidation stability, VI, and low-temperature fluidity.

5.9.1 Fuel Economy

In a review of the environmental benefits and impacts of engine lubrication, it is stated that optimization of three lubricant parameters—friction reduction, wear reduction and lubricant stability—will lead to positive environmental impacts. Low-viscosity lubricants, which may be made from synthetic or mineral oil blends, are less resistant to flow than conventional lubricants, a property that helps reduce friction and energy losses. Various studies have demonstrated fuel economy improvements ranging from 0.5 to 5 % with the use of low-viscosity engine lubricants and/or transmission lubricants. The cost savings realized with decreased fuel consumption (it is estimated that fuel savings outweigh the higher lubricant cost), complements a reduction in the amount of greenhouse gas and other contaminant emissions. Minimizing wear by efficient lubrication prolongs the useful life of an engine, thereby minimizing the consumption of non-renewable resources such as fossil energy and metal ores required for the manufacture and disposal of the machinery itself. In addition, wear of mechanical parts can cause the engine to operate less efficiently; thus wear reduction has a secondary benefit by reducing energy consumption throughout the operating lifetime of the engine. It is expected that lubricant developers will make increasing use of synthetic base fluids which permit optimization of engine performance through chemical design at the molecular level.

5.9.2 Extended Drainage/Reduced Oil Disposal

A factor influencing the environmental impact of lubricant oils is the stability or lifetime of the lubricant itself. Synthetic oils are generally more resistant to temperature changes, are less volatile than traditional oils, and are not as likely to oxidize in the engine environment. In the case of engine oils, the more stable a lubricant, the less is consumed. If a lubricant can be made to last twice as long, only half as much lubricant will be required, with corresponding reductions in the energy and material requirements for lubricant manufacture. The increased longevity of synthetic lubrication oils in the engine will also reduce the environmental impact of lubricant disposal.

5.9.3 Particulate Emissions Reduction

Synthetic lubricant oils may play a role in the reduction of engine exhaust emissions. Engine lubricating oil has been implicated as a significant parent material in the formation of mobile source particulate matter (PM) emissions. Published data suggest that various lubricant properties affect the composition of engine exhaust emissions, and that synthetic lubricating oils (PAOs) yield lower pollutant emissions. A recent study found that diesel engine emissions of nitrogen oxides (NO_x) and PM were 8 % and 19–24 % lower, respectively, with a full synthetic PAO-based oil compared to mineral-based oil. Several other studies report similar results, with PM emission reductions reported to range from 2 to 50 % of total PM mass. However, a few studies have reported conflicting results, with PM mass emissions increasing by up to 20 % in a diesel engine, and up to a factor of 3 in a spark-ignition engine. The California Air Resources Board is currently supporting a planned research project that will characterize the significance of lubricating oil in PM formation and determine whether lubricating oil can be formulated to reduce PM emissions from mobile sources.

As discussed earlier, as lubricant oils lubricate the engine, they accumulate PAHs. The implication is that the use of longer-life synthetic lubricants may generate particulate emissions with higher concentrations of PAHs (these compounds are generally bound to particulate matter). On the other hand, if the use of synthetic lubricants results in reduced particulate emissions overall (as discussed in the above paragraph), this might offset the potentially greater PAH levels associated with engine particulate emissions.

5.9.4 High Temperature Stability

The major part of piston deposits consists of carbon and complex carbonized organic components. This resinous binder acts to accumulate and aggregate wear debris, soot, and acids.

Although the composition of the organic phase of deposits is quite complex, it is well established that it originates from hydrocarbon oxidation processes. Piston deposits also comprise inorganic salts (e.g., salts of Ba, Mg, Ca, Zn, and P) derived from ash-providing additives. Highly volatile parts of the oil, either formed by degradation processes or resulting from the base oil manufacturing process, are readily lost from the engine by evaporation. This results in viscosity increase and reduced lubricant volume. The tendency towards viscosity increase, through loss of volatile components, can be assessed by the Noack test. An approach for reducing the deposit-forming tendency and lubricant viscosity increase is accomplished using antioxidant systems that inhibit deposit formation by moderating the detrimental effect of the primary oxidation products that would otherwise form deposits. When these preventative antioxidant systems are used in conjunction with detergent/dispersant additives, lubricant stabilization is achieved (Hamblin and Rohrbach 2001).

In very temperature engine environment, antioxidants interrupt the degradation process in different ways according to their chemical structure. The two major classes of antioxidants are primary and secondary anti-oxidants. Primary antioxidants readily form radicals that eliminate the 'harmful' radicals generated by the degradation process. The antioxidant radical is stabilized initially via an aromatic ring and it cannot initiate a radical chain. Hence they mainly act, as chain-breaking antioxidants, and they react rapidly with peroxy radicals. Secondary antioxidants react with hydro-peroxides, via a redox reaction to yield non-radical, non-reactive products, and are therefore frequently called hydro-peroxide decomposers. Secondary antioxidants are particularly useful in synergistic combinations with primary anti-oxidants for stabilizing lubricants against thermo-oxidation when operating under high-temperature conditions.

In addition to antioxidants, engine oils require detergents and dispersants to minimize deposit formation and sludge. Dispersants assist in the solubility of otherwise insoluble oxidation products and soot particles and inhibit agglomeration. It is thought that detergents adsorb on to the metal surfaces in the engine, forming a film that reduces the adhesion of deposits on these surfaces (Hamblin and Rohrbach 2001).

5.9.5 Bio-degradability of Synthetic Lubricating Oils

Biodegradation (biologically-mediated breakdown of a chemical to simpler molecules) represents a major means of removal of oils from soil and water. Because of the loss of motor oil to the soil and aquatic environments via engine leaks, spills, and illegal disposal, the biodegradability of lubricant oils is of ecological importance. Laboratory tests which were done indicated that synthetic ester lubricants are degraded more rapidly in soil and in aquatic systems than traditional mineral oil-based products. PAOs show higher biodegradability than mineral oils of equivalent viscosity because of their higher degree of hydrocarbon chain linearity.

Within a class of synthetic lubricants, the percent of material biodegraded within a prescribed time period can cover a large range, and different biodegradability tests can give different results for the same lubricant type. For example, biodegradability of PAOs can range from 20 to 80 % after 21 days using a "primary biodegradability test," which measures the initial transformation from the parent material. Using this same test, biodegradation of mineral-based oils ranged from 10 to 45 %. Rates and extents of biodegradation vary considerably between laboratory and field situations, largely due to the influence of factors such as temperature, the types and number of microbes, and the availability of oxygen and water. It should be noted that biodegradation tests are conducted on fresh lubricants, but biodegradability may be altered as a result of the accumulation of metals and other contaminants in crankcase oil during use. Tests have shown that used synthetic ester lubricants degrade more slowly than fresh lubricants, although they still biodegrade more rapidly than mineral oil.

5.10 Impact of Recycling Used Oil

To achieve maximum energy conservation and environmental benefit, it is preferable to re-refine used oils into regenerated base oils that can be blended into finished lube oil products compared to combustion for heating value recovery. A recent study found that re-refining used oils saves about 8 % of the energy content of the used oil compared to combusting the oil for heating purposes. As motor oil formulations transition to non-conventional lubricants (i.e., synthetics and other highly refined base oils), it is likely that the quality of the used oil pool available for recycling will improve. Using synthetic oils as feedstock for re-refining will, in turn, yield a better lubricating, more valuable re-refined base oil product. One re-refiner maintains that modern technologies must aim at recovering these partially and completely synthetic components in the "re-raffinates" (residues from extraction processes) to the greatest extent possible. Another argument for recycling synthetic lubricant base fluids is the fact that their manufacture involves relatively higher process energy requirements than mineral oils.

Re-refining oils can lead to additional environmental benefits because the toxic heavy metals (e.g., zinc, lead, cadmium, and chromium) are extracted from the used oil. These metal compounds are solidified and stabilized into asphalt flux, thereby posing minimal environmental risk. If used oils are combusted, however, metals in the flue gases can be released into the atmosphere unless they are captured by air pollution abatement equipment.

5.11 Synthetic Lubricants and the Future

Higher quality base stocks and synthetic oils are expected to increase their share of the market as a result of new industry requirements that demand better quality base oils. Globally, the demand for these oils is projected to grow by as much as 20 % annually from 2004 through 2020. Currently, synthetic and other highly refined lubricant base oils (referred to as non-conventional lube oils) make up less than 4 % of the total worldwide consumption. Some industry sources indicate that the high cost of PAOs has limited its market share to about 2 % of the total lube oil production.

Environmental forces, including emission standards and fuel economy requirements, have impacted finished lubricant properties such as sulfur levels, volatility and viscosity. Extended drain intervals appear to be popular with consumers willing to pay for higher priced oils for the convenience of fewer oil changes. Engine manufacturers are adopting increasingly stringent specifications, expanding the market for premium performance oils (Groups III and IV in Europe, and Groups II and III in North America). As more stringent engine oil specifications increase the demand for higher quality base oils, the increased availability of these premium oils in turn promotes the development of even tougher engine oil specifications. The industry anticipates that the availability of significant amounts of high quality base oils globally may eventually lead to global lubricant specifications.

Lubricants will continue to evolve towards products with higher purity, lower volatility and longer life. New technologies will enable the use of new feed stocks (such as natural gas) to produce Group III base oils with properties superior to PAOs. Feedstock prices for PAOs will continue to be relatively high, a factor that will likely limit PAO-based lubricants to smaller, specialized markets.

Already efforts are underway to develop a process that converts waste plastic (polyethylene and polyethylene terephthalate, or PET) to lubricating base oil. The process uses pyrolysis, where high-molecular-weight molecules are converted to lower-molecular weight molecules in the lubricant oil range. The product can be further converted to unconventional (synthetic) base oil quality. Waste plastic is readily available and inexpensive, and its diversion from the waste stream would reduce the growing environmental and political problems associated with landfill disposal.

5.12 Conclusion

Although synthetic lubricant oils are more costly than conventional mineral base oils, they provide the consumer with the convenience of extended oil change intervals. It is projected that the demand for higher quality base oils such as synthetics will continue to increase in light of increasingly stringent engine specifications. Ashless anti-oxidants can minimize deposit lacquer precursors, thereby preventing thermo-oxidative engine oil degradation, and they also enable the formulation of low-ash, high-performance engine oils. For high temperature diesel engine, use of synthetic base enhances the required long drain intervals required and the low viscosity requirements in a good lubricating engine oil. Under higher severity conditions, say 200 °C, the inherent stability for an optimized synthetic ester formulation is found to be approximately 5 times greater than that of the mineral oil formulation, while retaining biodegradability equivalent to that of the vegetable based formulation.

Revision Questions

1. *List advantages of synthetic lubricants over mineral oils.*
2. *What the drive behind thrust to adopt synthetics?*
3. *What would you consider a draw back in users moving onto synthetics?*
4. *Equipment builders have named synthetics lubricants of the future, why?*
5. *What environmental aspects are inherent in the use of synthetics.*

References

Denton JE (2007) A review of the potential human and environmental health impacts of synthetic motor oil, California Environmental Protection Agency Office of Environmental Health Hazard Assessment Integrated Risk Assessment Branch, April 2007

Girotti G, Raimondi A, Blengini AG, Fino D (2011) The contribution of lube additives to the life cycle impacts of fully formulated petroleum-based lubricants, Am J Appl Sci 8(11):1232–1240, Science Publications 2011

Hamblin PC, Rohrbach P (2001) Piston deposit control using metal-free additives, Lubrication science 14-1, 14(6), November 2001

Li H, Hsu SM, Wang JC (2010) Thermal-oxidative characteristics of some high-temperature diesel lubricants. Cummins Engine Co., Columbus, OH, USA

Murphy WR, Blain DA, Galiano-Roth AS (2002) Benefits of synthetic lubricants in industrial applications

Wills JG (2005) Lubrication fundamentals. Marcel Dekker Inc., New York, USA

Chapter 6
Eco-friendly Base Oils

Abstract The impacts of current lubricating oils on the environment have been the reason for an increasing move towards the use of environmentally safe lubricants. However, the development of a common biodegradable base stock that could replace conventional ones remains a big challenge. Even synthetic lubricants, whether synthetic hydrocarbons, or organic esters have problems associated with their use, despite the fact that they protect better, last longer and outperform their conventional mineral-based counterparts in certain applications. Future lubricant specifications in view of the demand for improved performance to meet stringent environmental regulations are the main drivers for new technological developments.

6.1 Introduction

The depletion of the current crude oil reserves, and issues related to conservation have brought about renewed interest in the use of bio-based materials. Emphasis on the development of renewable, biodegradable, and environmentally friendly lubricants has resulted in the widespread use of natural oils and fats. Vegetable oils are promising candidates as base fluid for eco-friendly lubricants because of their excellent lubricity, biodegradability, viscosity-temperature characteristics and low volatility. In view of agriculture based Indian economy, there is a great potential of producing vegetable oil based lubricants, which has ecological compatibility in addition to technical performance. However, suitability of the vegetable oils for a specific application either needs chemical modification or may be used as it is with additive blending route in order to get bases tocks as per specifications for a particular end use application (Srivastava and Sahai 2013).

The demand for biodegradable and environmentally friendly lubricants is on the increase, especially for use in ecologically sensitive areas. The environmental impact of large amounts of lubricants and industrial fluids has become an increasingly important issue, and the need for eco-friendly products has been

© Springer International Publishing Switzerland 2016
I. Madanhire and C. Mbohwa, *Mitigating Environmental Impact of Petroleum Lubricants*, DOI 10.1007/978-3-319-31358-0_6

the center of discussion. New applications are continually being found for these ecologically harmless lubricants, and for every successful application, the results obtained serve as pointers for further use of these products, underlining the need to develop novel biodegradable and ecologically harmless base stocks for future generations of lubricants. Synthetic fluids that are manufactured commercially are mostly ester-based poly alkylene glycols or vegetable-oil based fluids. Before ecological aspects became a consideration in lubricant development, ester oils were used in special lubricants such as base fluids for aviation turbine oils and components for fuel economy oils, compressor oils and other industrial lube applications. Environmentally acceptable lubricants can be formulated using synthetic polyol esters of a specific type of unconventional eco-friendly catalyst to come up with biodegradable oil formulations as base stocks.

Deliberate and accidental lubricant losses to the environment by means including evaporation, leakages, and spills have lead to major concerns regarding pollution and environmental health. About 5–10 million tons of petroleum products enter the environment every year, with 40 % of that representing spills, industrial and municipal waste, urban runoff, refinery processes, and condensation from marine engine exhaust. Thus, strict specifications on biodegradability, toxicity, occupational health and safety, and emissions have become mandatory in certain applications. The enactment of these specifications, along with uncertainty in the petroleum supply for political and economic reasons, has stimulated the search for alternative energy sources (Srivastava and Sahai 2013).

6.2 Bio-based Base Oil

Vegetable oils, a renewable resource, are finding their way into lubricants for industrial and transportation applications. Waste disposal is also of less concern for vegetable oil-based products because of their environment-friendly and nontoxic nature. Synthetic lubricant base oils are also available and offer improved stability and performance characteristics over refined petroleum oils, but at a premium price. Most of the biodegradable synthetic oils are esters that offer superior thermal and oxidative stability.

All continents have a great potential of producing edible and non edible tree borne oils, which remain untapped and can be used as potential source for vegetable oil based lubricants with an objective of ecological compatibility in addition to technical performance. Thus research and development are being undertaken to consider the replacement of non-renewable raw materials with bio-based materials in environment friendly lubricants selected based on their potential for combined environmental benefits and overall return to agriculture and contribute significantly to identify key research opportunities through which new bio lubricants might come to replace mineral oil based lubricants (Srivastava and Sahai 2013).

6.3 Lubricant Base Stocks

Lubricants are complex formulated products consisting of 70–90 % base stocks with the right physical characteristics, mixed with functional additives to optimize the physical properties in order to meet a series of performance specifications. The base stocks can be mineral, synthetic or re-refined apart from vegetable oils. Mineral oil bases are the most common. They consist predominantly of hydrocarbons but also contain some sulphur and nitrogen compounds with traces of a number of metals. Synthetic oils among others include poly-alphaolefins (PAO), poly-alkylene glycols (PAGs) and synthetic esters.

PAOs are petrochemical derived synthetic oils that most resemble mineral oils. Poly-alkylene glycols (PAGs) are polymers from petrochemical origin commonly made from ethylene oxide and propylene oxides. A major disadvantage of both PAOs and PAGs is their poor solubility with regard to additives. Since the additives themselves must also be biodegradable, this limits the additive types which can be used to formulate effective lubricants from them. Nowadays, some manufacturers are blending diesters with PAOs to form base oils which are biodegradable, have good solubility, resist oxidation, and have good temperature viscosity characteristics. Others are blending synthetic diesters with vegetable oils to provide similar results. Synthetic esters form a large group of products, which can be either from petrochemical or oleo chemical origin. Esters can be categorized as: monoesters, di-esters, phthalate esters, polyol esters and complex esters. Triacylglycerol structure of vegetable oil makes it an excellent candidate for potential use as base stock (Srivastava and Sahai 2013).

When using vegetable oils and esters to formulate lubricants, several specifications have to be considered. First technical specifications about suitability for the application are either as it is or with special attention to chemical modification to get base stocks with appropriate characteristics of the bio lubricants and secondly taking into consideration that both the raw materials and additive have to be as harmless as possible, and this certainly in accordance with local legal requirements in matter of health and environment.

Re-refined oil is used oil that undergoes an extensive refining process to remove contaminants to produce fresh base stock. These base stocks, irrespective of their origin, are then sold to blenders who add additive packages to produce lubricants.

6.4 Eco-labeling of Lubricants

Effort has been made to combine the environmental behavior and the technical properties of lubricants, as a number of countries have introduced "eco-labels", or "eco-logos" to try and give a sense of security to the users of environmentally compatible products. The eco-labeling system has issued criteria for lubricants. The parameters guided by the labeling criteria include toxicity, environmental impact (notably biodegradability), safety and renewability. Due to growing concern regarding the environmental impact and associated costs of lubricants, many lubricant companies have developed "environment friendly" lubricants or bio-lubricants.

In effect, bio-lubricants are nothing but the base stocks blended with additives and performance package aiming certain extent of rapid biodegradability and lower eco-toxicity as a deliberate and primary intention. Bio-lubricants are often but not necessarily based on vegetable oils. As the vegetable oils offer bio-degradability and low toxicity, it is obvious that during the formulation of a biodegradable and low toxicity fluid, the additive must also be biodegradable and have low toxicity. Bio-lubricants display a neutral CO_2 balance and easily decompose in nature. The currently marketed vegetable oil based lubricants are an ecologically sensible alternative with performance characteristics and quality comparable to that of their conventional petroleum based competitors.

6.5 Features of Good Bio-lubricants

Bio-degradable lubricants show less emission because of higher boiling temperature range of esters. They are totally free of aromatics, over 90 % bio-degradable and non water polluting. Also the oil mist and oil vapor reduction lead to less inhalation of oil mist into the lung. Bio-lubricants have better skin compatibility, less dermatological problems as well as high cleanliness at the working place. The high wetting tendency of polar esters leads to friction reduction with at least equal and often higher tool life. Higher viscosity index can be an advantage when designing lubricants for use over a wide temperature range. This can also result in lower viscosity classes for the same applications combined with easier heat transfer. Higher safety due to higher flashpoints at the same viscosities can be obtained leading to cost savings on account of less maintenance, man power, and storage and disposal costs.

6.6 Base Stocks from Vegetable Oils

Vegetable oils have a number of inherent qualities that give them advantages over petroleum oils as the feedstock for lubricants. Based on the fact that vegetable oils are derived from a renewable resource, they avoid the upstream pollution associated with petroleum extraction and refining in terms of usage. From a worker's safety perspective, plant-based lubricants are more attractive than their petroleum counterparts because of their relative low toxicity, high flash point and low volatile-organic compound (VOC) emissions.

The performance limitations of vegetable based lubricants stem from inherent properties of the vegetable oil base stocks rather than composition of additive package. Base stocks usually comprise nearly entirely predefine properties such as high biodegradability, low volatility, ideal cleanliness, high solvency for lubricant additives, miscibility with other types of system fluids, negligible effects on seals and elastomer and other less significant properties (e.g. density or heat conductivity).

Base stocks are also a major factor in determining oxidative stability, deposit forming tendencies, low temperature solidification, hydrolytic stability and viscometric properties. On the other hand, parameters like lubricity, anti-wear protection, load carrying capacity, corrosion (rust) prevention, acidity, ash content, color, foaming, de-emulsification (so called demulsibility), water rejection and a number of others are mostly dependent on the additives and impurities such as contaminants.

In this regard, when a fluid is considered for its suitability as a lubricant, first of all the base stock-dependent parameters are evaluated. In addition to biodegradability, the characteristics to be given attention are cleanliness (particle count), compatibility with mineral oil lubricants and homogeneity during long term storage, water content and acidity, viscosity, viscosity index, pour point, cloud point, cold storage, volatility, oxidative stability (for vegetable oils also iodine value), elastomer compatibility and possibly other properties, depending on intended application. Water rejection, demulsibility, corrosion protection, ash content and foaming could also be tested if contamination of the additive-free oil is suspected.

Vegetable oils clearly outperform mineral oils in terms of volatility or viscosity index. Many of the other properties are similar between the fluids or may be improved using additives. However, low resistance to oxidative degradation and poor low temperature properties are major issues for vegetable oils to be taken care of. It has been established that methylene interrupted poly unsaturation is the key factor causing low oxidative stability of vegetable oils. Low temperature properties to some degree can be characterized by determining the pour points. A similar test can also be used to determine the ability to remain liquid upon cold storage, which is often a concern in the case of vegetable-based lubricants. Low temperature properties of vegetable oils are inferior to those of synthetics or even mineral oil base stocks.

Four main vegetable oils dominate the industry accounting for about 82–85 % of worldwide vegetable oil production. Soybean oil is 31–35 %, palm oil 28–30 %, rapeseed oil 14–15 % and sunflower oil is 8–10 % of global production. A large portion of vegetable oil production is in developing countries such as India, Brazil, Argentina etc. (Srivastava and Sahai 2013).

6.7 Bio Lubricants Market

The market for bio-lubricants will be driven by environmental concerns, as well as by economics and performance issues. Vegetable oil based lubricants generally have lower performance characteristics than their petroleum based competitors, thus the initial market focus has been on "total loss" lubricants like chain bar and two stroke engine oils etc. taking advantage of the lower eco-toxicity and more favorable biodegradability characteristics of environment friendly lubricants. Despite the small current market, potential is large.

Each application area has unique performance requirements. To expand market opportunities, one market segment that justifies more research is the huge motor and gear oil market, which accounts for about half of the total lubricant market. This means that bio-products still have an enormous market potential. However, reality looks different. Although the bio-products are available on the market since at least five years, they have struggled from the beginning to compete with the well-established mineral oil-based products. Major technical performance barriers must be overcome to enable vegetable oils to be more widely competitive with mineral oils.

Price is a major barrier in the development of bio-lubricants as the oil seed prices are not competitive in comparison with the world market prices for many vegetable and mineral oils. However, due to instability in the price of mineral oils and the growing trade imbalances globally, use of indigenous renewable raw materials is an important logical step in view of severe predictions for the future. Bio-lubricants are generally between 1.5 to 5 times more expensive than mineral oils.

Economical and environmental balance needs to be performed in order to minimize the higher price that is, economic cost versus true cost in terms of their advantages and hidden cost savings on account of less maintenance, manpower, storage and disposal costs. This explains why the market share of vegetable based bio-lubricants is only about 3–4 % of the total use of lubricants and hydraulic fluids.

6.8 Make-up of Vegetable Oils

Vegetable oils consist primarily of long chains of carbon and hydrogen with terminal ends containing carbon, hydrogen and oxygen. This differentiates them from petroleum "hydrocarbons" which consist of chains and rings of carbon and hydrogen. Vegetable oils also differ from hydrocarbons as the carbons may or may not be "saturated" with hydrogen. The fact that vegetable oils make very good alternatives to petroleum hydrocarbons as the more saturated carbon behaves more like petroleum.

Vegetable oils in some cases outperform petroleum for example, bio-based motor oils in terms of friction reduction. However, they contain oxygen and some percentage of unsaturated carbon so they become more reactive and will polymerize, forming long chains of molecules. Most vegetable oils contain 18 carbon chains. The popular theory has long been that the longer the chain, the better the lubricant, but this may not be true in many cases. Longer chain acids like erucic acid (22 carbons) from rapeseed and crambe are not as good lubricating oil as canola (18 carbons). Canola also contains a small percentage of palmitic (16 carbon saturated) and stearic (18 carbon saturated) in its oil. The ability of these oils to aid lubrication is unknown because these fatty acids constitute small percentage of the oil. The use of palm oil in bio-lubricants requires a greater percentage of pour point depressant if used in temperate climates.

Design of environment friendly lubricants seeks the base stocks to meet commercial, technical and environmental needs. The challenge for researchers in this field will be to improve certain characteristics of vegetable oils without impairing their excellent tribological and environmentally relevant properties. This implies the utilization and sustainment of the natural chemistry of vegetable oils to a high extent. A preliminary chemical evaluation of base stocks should detect defects that may cause problems during in-use operation. In order to control the behavior during use, it is necessary to control the hydrolytic stability, control of the physicochemical and environmental characteristics during use, compatibility with materials and seals, corrosivity and wear and friction properties, apart from the two main issues of operational temperature limitations and oxidative instability. The selection of the type of typical physico-chemical and friction and wear test depends also on the final application of the oil.

Positively addressing these concerns has led to the development of new additive systems thus widening the choice of base stocks enabling even standard vegetable oil to meet the requirements for certain low temperature applications in addition to chemically modifying the vegetable base oil and genetically modifying the oilseed crop.

6.9 Additive Reformulation for Bio Lubricants

A lubricant consists of base oil and an additive package designed to improve performance characteristics such as oxidative stability, pour point and viscosity index. In the past, it was assumed that additives designed for mineral oil lubricants perform similarly when used with vegetable oils. This led to the production of technically inferior plant-based lubricants. Many of the industry's negative opinions of plant-based lubricants stem from these early products. More recently, companies have started reformulating and/or designing additives specifically for vegetable oils with successful results. Purely plant-based additive packages are now available which produce a full line of total loss lubricants and hydraulic fluids.

6.10 Chemical Modification of Base Oils

Some of the rapidly biodegradable lubricants are based on pure, unmodified vegetable oils. The triacylglycerol structure of vegetable oil makes it an excellent candidate for potential use as a base stock for lubricants and functional fluids.

Chemically, these are esters of glycerol and long-chain fatty acids (triglycerides). The alcohol component (glycerol) is the same in all vegetable oils. The fatty acid components are plant-specific, therefore, it varies and differs in chain length and number of double bonds, besides, functional groups may be present. Natural triglycerides are very rapidly biodegradable and are highly effective lubricants. However, their thermal, oxidation and hydrolytic stability are limited, their inferior low temperature behavior, and other tribo-chemical degrading processes that occur under severe conditions of temperature, pressure, shear stress, metal surface and environment also restrict their use as lube base stocks in its natural form. Therefore, pure vegetable oils are used only in applications with low thermal stress. These include total loss applications like mold release and chain saw oils.

Chemical modification is necessary to improve these performance limitations with the focus on eliminating bis allelic hydrogen functionalities in methylene interrupted poly unsaturation and optimal extent of structural alteration for improved low temperature performance. Chemical modification of vegetable oils is an attractive way of solving these problems. Bio-lubricants formulated from plant oils should have the following advantages derived from the chemistry of the base stock:

- Higher lubricity leading to lower friction losses, yielding more power, and better fuel economy.
- Lower volatility resulting in decreased exhaust emissions.
- Higher viscosity indices.
- Higher shear stability.
- Higher detergency eliminating the need for detergent additives.
- Higher dispersancy.
- Rapid biodegradation and hence decreased environmental/toxicological hazards.

Modifications of the carboxyl group: One of the most important modifications of the carboxyl group of the fatty acid chain, performed on large scale is the transesterification of the glycerol esters or the esterification of the fatty acids obtained by cleavage of such esters, which are normally catalyzed with acidic or basic catalysts. Typical homogeneous catalysts are p-toluene sulphonic acid, phosphoric acid, sulphuric acid, sodium hydroxide, sodium ethoxide and sodium methoxide.

Transesterification (synthesis of esters): The major components of vegetable oils and animal fats are Triglycerides. To obtain ester, the vegetable oil or animal fat is subjected to a chemical reaction termed transesterification . A number of transesterified and alkylated derivatives have been synthesized from available vegetable oils. It has been found that several esters of fatty acids of these vegetable oils have a high natural viscosity index, low pour points, and high thermo oxidative stability, and can meet the requirements as base fluid components for energy-efficient, eco-friendly, long-drain interval, multigrade oils. These oils have markedly lower viscosities at 40 °C, higher load-carrying characteristics, and lower friction coefficients than the base fluids of currently marketed multigrade oils. A 50 % blend with hydrorefined hydrocarbon oils could prove highly viable. It clearly establishes the potential for utilizing these esters, either alone, or in combination with mineral oils, for formulating cost-effective high-performance, energy-efficient, and environmentally friendly lubricants. Biodegradable organic polyesters derived from the transesterification/esterification of vegetable oils and branched neopolyols such as trimethylolpropane (TMP) and poly esters have been developed for various applications.

Modification of fatty acid chain: In the field of lubricant base fluids, reactions at the double bonds are used to increase the temperature and hydrolytic stability but somewhat lowers the viscosity index. Viscosity index increases with linearity whereas low pour characteristics gradually deteriorates. Furthermore, the polarity can be altered by the introduction of hetero-atoms such as oxygen or nitrogen. Thus, there is a great potential for economic exploitation of vegetable oils and fats to have a new product line of bio-lubricants.

Selective hydrogenation: This process primarily serves to improve the melting point as well as the ageing behavior. Selective hydrogenation, in which the fatty acid residue is not fully saturated, is of greatest interest in the area of lubricant chemistry. The stability of the oils can be significantly improved by selective hydrogenation of easily oxidizable compounds into more stable components. It transforms the multiple unsaturated fatty acids into single unsaturated fatty acids without increasing the saturation, necessary to avoid deterioration in low-temperature behavior.

Epoxidation: Epoxidation is one of the most important double bond addition reactions. In the case of unsaturated fatty acid esters, it is often performed in situ using the per formic acid method. This process is industrially performed on a large scale. At present the vegetable oil epoxides are used in polyvinyl chloride (PVC) and stabilizers. Furthermore, they are also used to improve the lubricity in lubricants. Due to their good lubricity and high oxidation stability in comparison to rapeseed oil, pure epoxidised rapeseed oil can also be used as a lubricant base fluid. The chemically modified base fluids exhibit superior oxidation stability in comparison with unmodified vegetable oils. These base fluids in combination with suitable additives exhibit equivalent oxidation stability compared with mineral oil-based formulations.

Estolides of oleic acid and saturated fatty acids: Estolides are a class of esters-based on vegetable oils and are synthesized by the formation of a carbocation at the site of unsaturation. This carbocation can undergo nucleophilic attack by other fatty acids, with or without carbocation migration along the length of the chain, to form an ester linkage. Estolides were developed to overcome some of the short-falls associated with vegetable oils, such as poor thermal oxidative stability and poor low temperature properties. Some deficiencies can be improved with the use of additives but usually at the expense of biodegradability, toxicity and cost.

Dimerization: Another technologically feasible modification of the double bonds of unsaturated fatty acids is their dimerisation and oligomerisation involving two or more fatty acid molecules being attached to the residual alkyls. C18 fatty acids with one or more double bonds react with each other at temperatures of about 210–250 °C in the presence of layered aluminosilicate catalysts (e.g. montmorillonite) forming a complex mixture of C_{36} dicarboxylic acids (dimeric fatty acids), C_{54} trimer fatty acids and C_{18} monomer fatty acids. Hydrogenation of the double bonds in those compounds results in a solid acids whose branching points are mainly concentrated in the centre of the molecule.

Branching of fatty acids: Branched fatty acids are interesting base fluids for lubricants because of their extraordinary physical features. For example, the pour point of fatty acids and their derivatives is significantly lowered by branching. By adding sterically hindered branches, hydrolytic stability can also be increased. Their low pour point, low viscosity, high chemical stability and high flashpoint make saturated, branched fatty acids highly desirable base fluids in the lubricants industry. Branching of the fatty acids residues is initiated by C–C and C–O linkages.

Modified cultivations of natural vegetable oils: Apart from a chemical modification, the oil industry is working together with plant cultivators to develop new oils whose fatty acid make-up is better suited for the demands of industrial applications as the industry is highly interested in tailor-made raw materials. Such natural structural uniformity of vegetable raw materials is utilizable input in the chemical processing. However, oleic acid or oleic acid esters still are the starting point of the chemical modifications. More than 90 % technical oleic acid content would produce much fewer by-products than with present raw materials resulting in more efficient reaction and purer end-products. One cultivation success is high oleic sunflower oil (HOSO). The extremely good physical and chemical properties of HOSO may lead to this vegetable oil being used in the non-food sector on a large scale. Even unmodified, it is suitable as base fluid for various biodegradable lubricants.

6.11 Synthetic Base Oil Synthesis

Esters are normally synthesized using catalysts based on PTS, Ni, Cu, Fe, V, CO, and Sn, Cu and Cr oxides, alkoxy zirconate, and hetero-poly acids. In these processes the catalysts are used for once-through application, have disposal problems, yield base oils with significant acidity, and are not eco-friendly. The use of the unconventional eco-friendly catalysts gives the products (esters) negligible free acidity.

Physico-chemical characteristics of the synthesized polyol ester lubricant base stocks have been shown to be of limited adverse impact on the environment. The presence of an ester group implies polarity, which is reflected in the vapor pressure, lubricity and solvency. Thus, in comparison with mineral oils, many esters have technical advantages. These include a naturally high viscosity index, good low-temperature properties (good cold flow), pour points of $<-26\,°C$, low evaporation losses, good thermal stability, i.e., high-temperature properties, hydrolytic stability, good anti-friction and anti-wear characteristics, high flash points ($>260\,°C$), seal compatibility, and biodegradability. Another important feature of these base fluids is their biodegradability, which is normally in the range 80–90 %, and the lubricity characteristics of these products are better than those of comparable mineral oil base stocks. From analysis of the physico-chemical test data, it was found that these products have a potential application as base stocks for even automotive transmission fluids (ATFs).

6.12 Bio-degradable Lubricants

Synthetic are used, for example, in power steering units and hydraulic systems, providing good lubricating protection. They have better performance and a wider temperature range of operation than do commercial petroleum fluids. In transmissions usually operate under severe conditions. In the field of lubrication it is difficult to formulate a product that will meet all the requirements for a particular use. Formulated oils provide cooling, lubrication, and rust protection for moving parts. A finished bio-degradable lubricant can be produced using an appropriate base stock and suitable additives. Certain features of these synthetic polyol esters are comparable to vegetable oils, but their better thermal and oxidative stability and lower pour points can usefully extend the scope of their successful application in lubricating oil formulations. Since a finished lubricant normally consists of greater than 90 % base oil, the polyol base stocks are potentially biodegradable.

6.13 Conclusion

In view of stringent emission norms and environmental compliance, research activities in the area of development of renewable base stocks for lubricant formulation are rapidly increasing. Due to specific structure of vegetable oils, they combine good boundary friction lubricity and general wear protection along with stable viscosity-temperature behavior and very low evaporation. In addition, they have particularly good tribological properties apart from excellent biodegradability and low toxicity. They are excellent raw materials for the formulation of bio lubricants. All their highly positive physical features are countered by a few major limitations, the most important of which is the inadequate ageing resistance of these products. As a result, they are only suitable for circulation lubricant systems. Vegetable oil-based products are perfect for total loss applications in which the lubricant inevitably enters the environmental cycle. The resulting high-performance lubricants are technologically much better than mineral oils. The decisive factors for this development is environmental protection and, to a lesser extent, technical requirements. However, emergence of new oils will always be slow in their initial development due to small volumes and the conservative nature of the industry.

Chemically modified renewable base stocks offer great potential for the development of bio-lubricants. Important modifications relate to the carboxyl group of fatty acids, e.g. esterification and fatty acid chain. In each case, the overriding objective is to realize application-relevant properties. Successes in this area reduce the dependence on petrochemical raw materials and create new synthesis processes. The challenge for researchers in the field of bio-lubricants will be to improve certain characteristics of vegetable oils without impairing their excellent tribological and environmentally relevant properties such as bio synthetic lubricants.

Synthetic polyol esters have a good potential for use in biodegradable lubricating oil formulations. The use of the unconventional eco-friendly catalysts in the synthesis of the products (esters) is beneficial as compared to conventional catalysts, offering advantages in terms of reuse (of the catalyst), free acidity, shorter reaction time, and other benefits, and the products, in the present study a formulation for automotive transmission fluids.

Revision Questions

1. *What short comings do ordinary ester based synthetics lubricants bases have?*
2. *How are the same overcome in polyol esters?*
3. *How do the polyol esters compare to the vegetable on the impacts on environment?*
4. *Why would good base oils such as polyol esters need commercial additives?*
5. *Give a typical total loss application where bio-degradable oil in a must.*

References

Srivastava A, Sahai P (2013) Vegetable oils as lube base stocks: a review, Afr J Biotechnol 12(9):880–891. http://www.academicjournals.org/AJB. doi:10.5897/AJB12.2823

Chapter 7
Development of Biodegradable Lubricants

Abstract A significant amount of lubricating oils can enter the environment, producing contamination of both soil and water. Pressure has been put on lubricant producers and consumers to spur them to seek a solution involving less environmentally harmful lubricants to mitigate the effect of disposing lubricants without control, polluting soil and water. Issues related to green pressures and conservation have brought about renewed interest in the use of bio-based materials. Emphasis on the development of renewable, biodegradable, and environmentally friendly industrial fluids, such as lubricants, has resulted in the widespread use of natural oils and fats for non-edible purposes. This chapter reviews the available literature and recently published data related to bio-based raw materials and the chemical modifications of raw materials. Additionally, it analyzes the impacts and benefits of the use of bio-based raw materials as functional fluids or bio-lubricants.

7.1 Introduction

It has long been known that environmental pollution caused by conventional mineral lubricants is significant because of their low biodegradability and even eco-toxicity. The increased public attention and awareness of protection of the environment has stimulated the lubricant industry to develop and introduce lubricants that show greater compatibility with the environment. New innovations in lubricant technology are increasingly concentrating upon and responding to the needs of the environment. The ecologically responsive technology and lubricant design seek to meet both performance and environmental needs to harmonize the technical performance with ecological requirements. Bio-lubricants are more attractive than their petroleum counterparts because of their high biodegradability and relatively low toxicity. The key issue in formulating biodegradable lubricants is the choice of reliable base oils and suitable performance additives. The impact of base oil should be considered before the rest of the formulation is considered.

© Springer International Publishing Switzerland 2016
I. Madanhire and C. Mbohwa, *Mitigating Environmental Impact
of Petroleum Lubricants*, DOI 10.1007/978-3-319-31358-0_7

Of late, many base fluids such as vegetable oils, synthetic esters and low viscosity poly-alpha-olefins have found a practical application in formulating biodegradable lubricants, mainly because of their excellent bio-degradability. Mineral based petroleum lubricants, which became available in the early 1900s, for many reasons still dominate the lubricant market, and continue to play an important role in future lubrication applications. While the poly-alphaolefins (PAOs) because of their superior technical performance in areas such as viscosity–temperature characteristic and fluidity at low temperature, have increasing promise as excellent lubricant base stock alternative. Apparently, both the mineral oils and high-molecular-weight PAOs are unreadily biodegradable. Therefore effort to enhance their biodegradability is of important practical significance from both technical and environmental points of view (Boshui et al. 2008).

The term bio-lubricants applies to all lubricants, which are both rapidly biodegradable and non-toxic to humans and other living organisms, especially in aquatic environments. Biodegradability provides an indication of the persistence of the substance in the environment and is the yardstick for assessing the eco-friendliness of substances. Scientists are discovering economical and safe ways to improve the properties of bio-lubricants, such as increasing their poor oxidative stability and decreasing high pour points. "Green" bio-lubricants must be used for all applications where there is an environmental risk.

Bio-lubricants for industrial and maintenance application, by contrast, may lead to a decrease in costs and to guarantee competitiveness. It is thus to be expected that environmentally sound applications of lubricants will be of great interest to all manufacturing companies. By the use of bio-lubricants, it is possible to reduce the use of petroleum based lubricants both in industrial and maintenance applications and also cut down the serious environmental problems caused (Willing 2001).

7.2 Drive for Environmental Compatible Lubricating Fluids

It is estimated that about 50 % of all the oil used ends up in the environment. Petroleum based lubricants, which are the leading type of base oil used in this industry, are poorly degradable and represent an environmental hazard when released. This represents a strong incentive to provide lubricants that are biodegradable. In addition, the rapid increase in the price of petroleum products in recent years, the increased dependence on offshore sources, the declining rate of production from older domestic oil fields and the decrease in the rate of finding new reserves has prompted governments and individuals to press for renewable products as replacements for petroleum products where practical. The bio-lubricant industry is growing based on these pressures, environmental concerns and sustainability. Bio-lubricants may be defined generally as materials that are based on biodegradable and renewable base stocks. However, this definition is not universally accepted. In some areas, only biodegradability is considered in the definition. For present purposes, bio-lubricants will be defined as biodegradable and renewable materials. Bio-lubricants do not have to be composed entirely of vegetable oil base stocks. They can be products derived from renewable oils, such as the fatty acids from fats and oils, reacted with synthetic alcohols or polyols to produce esters that can be considered bio-lubricants. Also, the natural vegetable oils can be treated to produce a modified product that is still biodegradable and renewable.

Bio-lubricants based on vegetable oils have to overcome their inherent instability based on the presence of poly-unsaturated products in the natural oil to compete with products based on mineral oils. Great strides have been made since the original 1997 study, and soyabean based oil bases have been improved by chemical transformation, formulation and improved additive technology. In spite of the improvements in the performance of bio-lubricants, the market for these products has been slow to develop. The reasons for this are price and the lack of regulatory pressures to change. Bio-lubricant products are generally more expensive than their mineral oil counterparts, with some notable exceptions, so without regulatory pressures it is difficult to convince a user to change from what they know to be acceptable performance from traditional mineral oil based products (Bremmer and Plonsker 2008).

7.3 Application of Bio-lubricants

Lubricants are used in various industrial and maintenance applications throughout the world. Some of the important applications are; Industrial Oils used for industrial purpose like machine oils, compressor oils, metal working fluids and hydraulic oils etc. Bio-lubricants provide significant advantages as an alternative lubricant for industrial and maintenance applications due to their superior inherent qualities. Bio-lubricants due to their environmental benefits enable their use in sensitive environments and provide pollution prevention. Bio-lubricants have capability of being utilized in various industrial and maintenance applications such as (Jaina and Suhane 2013):

Hydraulic fluids: Hydraulic fluids or hydraulic liquids are used to transmit power in hydraulic machinery. Mainly hydraulic fluids are based on petroleum oil. Petroleum oil can pollute drinking water, with the agricultural and mining industries being major consumers of hydraulic fluid. Bio based hydraulic fluids are used in environmentally sensitive applications when there is the risk of an oil spill. Vegetable oils are used as base stocks for fluids where biodegradability is considered important. Some of the significant advantages of using biodegradable hydraulic fluids are viscosity for film maintenance, low temperature fluidity, cleanliness and filterability, anti-wear characteristics, corrosion control, adequate viscosity and viscosity index, shear stability, low volatility, proper viscosity to minimize internal leakage, high viscosity index.

Metal working fluid: Metalworking fluids reduce heat and friction and also remove metal particles in industrial machining and grinding operations. Bio-based metal working fluids have excellent lubricity and viscosity versus temperature characteristics, better thin film strength, less smoke and risk of fire. The use of bio-lubricant can fulfill the requirement for energy independency and a safe environment.

Metal forming: Bio-lubricants are used in metal forming operations to separate work piece and tool surfaces, reduce interface friction, ease metal flow in order to produce sound components and increase tool life.

Cutting fluids: Bio based cutting fluids are used for improving tool life, reduce thermal deformation, improving surface finish, reduce environmental impact and are safe and convenient for use in a wide variety of manufacturing applications (Jaina and Suhane 2013).

Grease: Semisolid lubricants are applied to mechanisms that can only be lubricated infrequently and where lubricating oil would not stay in position. Grease also acts as sealants to prevent entry of water and incompressible materials. Bearings lubricated with grease have greater frictional characteristics due to their high viscosity. Rape seed oil based grease was found to be rapidly biodegradable as a general grease containing a calcium soap for total-loss applications, for operations between -20 to $+70\ °C$.

Concrete Mould Release Agents: Bio-lubricant prevents freshly poured concrete from sticking to its mould or form work and thereby facilitates removal of the formwork once it has cured. It can be used for environmentally sensitive areas.

Chain Saw Oils: Specially designed bar and chain combinations as a tool for use in chain saw art. Bio based chain saw oils have excellent lubricity and good ageing stability.

Gear oils: Biodegradable and environmentally friendly, high-performance gear oils provide excellent protection against micro-pitting, eliminate or reduce smoke, improve tool life, removal of foreign or wear particles (from critical contact areas of gear tooth surfaces) and corrosion prevention.

Gearing Applications: Bio based oil in gearing applications increase tool life. At normal feeds and speeds tool life increases by up to 100 %. Faster machining processes and lesser tool change can lead to productivity increases of 20 % or even more. Better tool condition and improved surface finishes improves part quality. Processes like grinding, hobbing, shaving, shaping and broaching can be benefited by the inherent characteristics found in bio-lubricants.

Gear grinding: Bio-lubricants provide important meliorations in gear grinding like reduced friction which improves dressing efficiency and improves dresser life, the heat produced allows for faster grinding with lesser chance of burning the part, and it also provide improved surface finish, grinding wheels retain their size which extend wheel life, cutting edges is improved to prevent premature cratering which increase tool life.

Grinding operations: Bio based coolants can benefit greatly in grinding operations; reduce friction causing abrasive grains to stay sharp longer, allowing much longer time between wheel dresses so wheel life and accuracy is improved, reduction in heat generated in most grinding operations this allows feed rates to be increased without introduction of excessive heat into the part.

Grinding operations: Bio based coolants can benefit greatly in grinding operations; reduce friction causing abrasive grains to stay sharp longer, allowing much longer time between wheel dresses so wheel life and accuracy is improved, reduction in heat generated in most grinding operations this allows feed rates to be increased without introduction of excessive heat into the part.

Drilling fluid: Boreholes into the earth are assisted by drilling fluid, used while drilling oil and natural gas wells and on exploration drilling rigs, also used for simple boreholes (water wells). Drilling fluids keep the drill bit cool, provide hydrostatic pressure to prevent formation fluids from entering into the well bore, avoid formation damage and to limit corrosion. A drilling fluid produced from bio based ingredients that are able to withstand the harsh and challenging down hole environment conditions, outperforming the traditional fluids and the residual fluid and cuttings are able to meet all environmental standards to return to the environment, thus able to offer a much cleaner and safer approach.

Slide way oils: Biodegradable and environmentally friendly slide way oils have very good friction coefficients, eliminate or reduce smoke, improve tool life, good EP and anti-wear performance.

Chain lubricants: Bio based tacky lubricant for chains and other total-loss applications.

Machining process: Machining processes in which a piece of raw material is cut into a desired final shape and size by a controlled material-removal process. Application of bio-lubricants on some of the common machining process is shown in the Fig. 7.1.

Water soluble coolants: In machining bio-lubricant offer features that can lead to cost savings, quality improvements, improved lubricity for better tool life, better surface finish, reduce friction, reduce heat and increased productivity. Bio based coolants are easy to maintain due to excellent sump stability and reduced toxicity. No inflammation of the skin is experienced by operators.

Bio based coolants work well on all metals like stainless steels, alloy, and tool steels, as they improve tool life, long sump life, and low maintenance. Bio lubricants and coolants are easily cleaned from parts and machines with soap and water. Residue left on steel parts acts as an effective in-process rust preventative non-staining on most metals, including brass and most aluminum alloys. In terms of health and safety, bio lubricants are not ecologically hazardous; OSHA (Occupational Safety and Health Administration) limits for bio based vapors are much higher than for petroleum based vapors. No skin problem as the operator's skin does not dry out after contact with these oils. Vegetable oil is far less toxic than petroleum based lubricant, glycols and synthetic oils. Vegetable oil has much higher flash point (approximately 275–290 °C) which reduce the risk of accidental ignition, eliminating all injuries, unsafe practices, occupational illnesses and incidents of environmental pollution (Jaina and Suhane 2013).

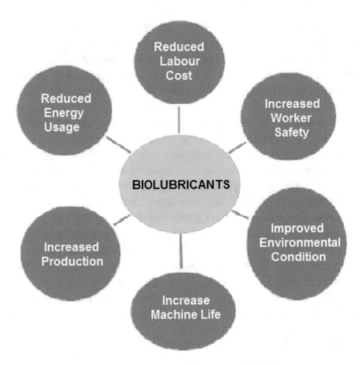

Fig. 7.1 Some benefits of bio lubricants. *Source* Salimon et al. (2010)

7.4 Vegetable-Based Bio-lubricants and the Environment

Bio lubricant is an alternative lubricant different from mineral oil lubricant as it is prepared from non-conventional energy resources and is non toxic, bio-degradable and eco friendly. There is an increasing concern for environmental pollution from excessive petroleum based lubricants use and their disposal especially in lost lubrication, military applications, and in outdoor activities such as forestry, mining, railroads, dredging, fishing and agriculture hydraulic systems. Vegetable oils with high oleic content are considered to be potential candidates to substitute conventional mineral oil-based lubricating oils and synthetic esters. Vegetable oils are preferred over synthetic fluids because they are renewable resources and cheaper. Furthermore, vegetable oils lubricants are biodegradable and non-toxic, unlike conventional mineral-based oils. They have very low volatility due to the high molecular weight of the triacylglycerol molecule and have narrow range of viscosity changes with temperature. Polar ester groups are able to adhere to metal surfaces, and therefore, possess good boundary lubrication properties. In addition, vegetable oils have high solubilizing power for polar contaminants and additive molecules. Biodegradable greases are good candidates for food processing and water-management machinery (Chauhan and Chhibber 2013).

Vegetable oil base stocks have the following advantages at ambient conditions when compared to mineral oil base stocks: higher viscosity index, lower evaporation loss and a potential to enhance lubricity, which could lead to improved energy efficiency. The enhanced lubricity of vegetable oils is mainly attributed to their chemical composition of typically 80–95 % fatty acids. However, it is well known that vegetable oils have poor thermal, oxidative and hydrolytic stability. Mineral oil, on the other hand, typically contains saturated aliphatic compounds, namely paraffinic and naphathenic, and small amounts of aromatics, which are more stable at elevated temperatures.

The present emphasis on conservation has brought about renewed interest in the use of these "natural oils" for non-edible purposes. The sources of natural oils and fats come from various plants and animals-based raw materials (e.g., soya bean, palm, tallow, lard). Plant oils are superior in terms of bio-degradability, especially when compared to mineral oils. Attention has been focused on technologies that incorporate plant oils as bio-fuels and industrial lubricants, due to the fact that they are renewable and non-toxic.

The application of plant oils and animal fats for industrial purposes, specifically as lubricants, has been in practice for many years. Environmental and economic reasons lead to the utilization of plant oils and animal fats, or used oils and fats after their appropriate chemical modification. Plant oil-based lubricants and derivatives have excellent lubricity and bio-degradability, for which they are being investigated as a base stock for lubricants and functional fluids (Salimon et al. 2010).

Strong environmental concerns and growing regulations over contamination and pollution in the environment have increased the need for renewable and bio-degradable lubricants. Accelerating research and development in this area has also been driven by public demand, industrial concern, and government agencies. Better ways to protect the eco-system, reduce the negative impact of spills or leakage of lubricants must be outlined. Many terms are used for the classification of lubricants and include products that are environmentally friendly, environmentally acceptable, biodegradable, non-toxic, etc.

About 45 % of all used lubricants return to the environment with altered physical properties and appearances. These lubricants included those used in frictional loss lubrication.

Lubricants that remain in the environment also include those used in circulation systems, which are not collected and disposed. In addition, leaked lubricants and those remaining in filters or containers have to be taken into account. Altogether, the environment is exposed to thousands of tons of lubricants annually, if lost lubricants and undefined lubricants are accounted for as returning to the environment.

The production, application, and disposal of lubricants have to meet the requirements for the best possible protection of the environment and of living beings in particular. Most often, health hazards to humans are derived from indirect routes through the environment. For all cases of direct contact between lubricants and human beings, compatibility has to be verified. All measures have to be taken to keep the impairment of the environment at the lowest possible level. In evaluating acceptable detrimental effects upon the environment, the benefit of lubricants, such as their performance or economic properties, must be considered and weighed against the risks associated with these lubricants.

Bio lubricants are increasingly becoming the best solution for the toxic effects that are caused by petroleum based lubricants on our ecosystem. No environmental problems will be caused from hydraulic leaks, no injuries at the job so no lost in work which will decrease maintenance between oil changes more production increases profits. Employees no longer have to experience the inflammation of the skin caused by petroleum based lubricants. Furthermore bio lubricants in industry provides significant advantages as shown in Fig. 7.1 due to their super high inherent characteristics. Environmental factors are gaining importance in our society. A large amount of petroleum based lubricants pollute the environment during or after use, mostly from spills and industrial processes. As environmental concerns gaining interest the bio lubricant industry is growing and various countries are restricting the use of petroleum based lubricants in applications where lubricants can get in contact with soil and water (Jaina and Suhane 2013).

In summary, capability of bio lubricants as alternative lubricant for industrial and maintenance applications include lower toxicity, good lubricating properties, high viscosity index, high ignition temperature, increased equipment service life, high load carrying abilities, good anti-wear character, excellent coefficient of friction, natural multi grade properties, low evaporation rates—low emissions to atmosphere, and rapid biodegradability. Industries can reduce tool costs and improving product quality in a safer environment by switching to bio lubricants. Bio lubricant reduces costs and increases competitiveness.

7.5 Biodegradable Base Stocks and the Environment

A lubricant is a substance (often a liquid) introduced between two moving surfaces to reduce the friction between them, improving efficiency and reducing wear. Lubricants dissolve or transport foreign particles and distribute heat. Some biolubricants also contain small amounts of additives. Plant oils or synthetic liquids such as hydrogenated poly-olefins, esters, silicone, and fluorocarbons are used as base oils. Additives deliver reduced friction and wear, increased viscosity, improved viscosity index, resistance to corrosion and oxidation, aging or contamination, etc. In addition to industrial applications, lubricants are used for many other purposes. Other uses of lubricants include bio-medical applications (e.g., lubricants for artificial joints) and other personal purposes. In an attempt to classify lubricants according environmental risk, many different terms have been established. It is useful to rank the terms related to environment risk, as in Table 7.1.

According to the above categorization, no lubricant can be regarded as environmentally friendly, because this term implies an improvement to the environmental conditions. One has to be content with the fact that the

Table 7.1 Common terms related to the environment

Term	Effect on environment
Friendly	Improves the environment
Neutral	Unimportant, harmless
Threshold of perception	–
Sociable	–
Start of legal regulations	–
Annoying	Disagreeable, unpleasant, impairing
Irksome	Troublesome, inconvenient
Limit of burdening	–
Endangering	Excessive, un-imputable
Harmful	Dangerous, irreversible effects

lubricant is environmentally acceptable and that it affects the environment to a less pronounced degree.

Furthermore, biodegradability is the most important aspect with regard to the environmental fate of a substance. Primary degradation is the first step in the breakdown of a substance and involves the disappearance of the original molecule. However, the determination of the ultimate degradability or the mineralization of substances to CO_2, H_2O, and the formation of biomass is important. Ultimate biodegradability guarantees the safe reintegration of the organic material in the natural carbon cycle and is important for its environmental classification. Biodegradability depends more on the chemical structure of the lubricant than on its water solubility.

On the other hand, vegetable oils have poor oxidative stability primarily due to the presence of bis allylic protons and are highly susceptible to radical attack and subsequently undergo oxidative degradation to form polar oxy compounds. This phenomena result in insoluble deposits and increases in oil acidity and viscosity. Vegetable oil also shows poor corrosion protection. The presence of ester functionality renders these oils susceptible to hydrolytic breakdown. Therefore, contamination with water in the form of emulsion must be prevented at every stage. These physical and chemical properties can be improved either using genetically modified oils or chemically modified oil with suitable combination of additives.

Non edible oils can be converted into bio lubricant by chemical modification which includes acylation, epoxidation, partial hydrogenation and trans-esterification. Trans-esterification process was found to be most viable process. Trans-esterification is the process of using an alcohol (e.g. methanol, ethanol, or ethyl hexanol), in the presence of catalyst, such as sodium hydroxide or potassium hydroxide, to break the molecule of the raw renewable oil chemically into methyl or ethyl esters of the renewable oil, with glycerol as a byproduct. Trans-esterification is a chemical reaction that aims at substituting the glycerol of the glycerides with three molecules of monoalcohols such as methanol thus leading to three molecules of methyl ester of vegetable oil.

Concerns over the discharge and accumulation of lubricants and fuels on land, water and air posing serious hazards to health and deleterious effects on the environment led to the framing of increasing stringent state policies discouraging the use of conventional petroleum based lubricants in several applications such as total loss lubricants, industrial lubricants for food processing, water-management machinery, two-stroke engine lubricants, etc. and encouraging their replacement with rapidly biodegradable lubricants of low toxicity. There are moves to replace mineral oil based lubricants in high powered diesel engine vehicles with low evaporation loss ester based lubricants in order to reduce particulate emissions which pose serious respiratory problems in large cities (Chauhan and Chhibber 2013).

7.6   Basic Eco-toxicological Properties of Bio-lubricants

The term bio-lubricants applies to all lubricants that are both rapidly biodegradable and non-toxic to humans and aquatic environments. A bio-lubricant may be plant oil-based (e.g., rapeseed oil) or derived from synthetic esters manufactured from modified renewable oils or from mineral oil-based products.

Biodegradation is the process by which organic substances are broken down by the enzymes produced by living organisms. The term is often used in relation to ecology, waste management, and environmental remediation (bio-remediation). Organic material can be degraded aerobically, with oxygen or anaerobically, without oxygen. A term related to biodegradation is bio-mineralization, in which organic matter is converted into minerals. By definition, biodegradation is the chemical transformation of a substance by organisms or their enzymes. Table 7.2 summarizes some of the benefits of bio-lubricants.

Table 7.2 Benefits of biodegradable lubricants

Aspect	Aspect details
Less emission	Due to the higher boiling temperatures of esters. Native triacylglycerol leads to partly gummy structures at high temperature and can accumulate acroleins, which are irritating
Totally free of aromatics	Over 90 % biodegradable oils, non-water polluting
Oil mist and oil vapor reduction	Leads to less inhalation of oil mist into the lungs
Better skin compatibility	Less dermatological effects
High cleanliness at the work place	
Equal and often higher tool life	Due to a higher wetting tendency of polar esters, which leads to a reduction in friction
Higher viscosity index	Viscosity does not vary with temperature as much as mineral oil. This can be an advantage when designing lubricants for use over a wide temperature range
Higher safety on shop floor	Higher flashpoints at the same viscosity
Cost savings	On account of less maintenance, man power, storage, and disposal costs

7.7 Development of High-Performance Industrial Bio-lubricants

Plant oils have different unique properties compared to mineral oils, due to their unique chemical structure. Plant oils have a greater ability to lubricate and higher viscosity indices. Superior anticorrosion properties are observed in vegetable oils and are induced by a greater affinity for metal surfaces. High flash points over 300 °C classify vegetable oils as nonflammable liquids. However, the applicability of vegetable oils in lubrication is partly limited, as these oils tend to show low oxidative stability and higher melting points. Chemical modification of vegetable oils is an attractive way of solving these problems. Bio-lubricants formulated from plant oils should have the following advantages derived from the chemistry of the base stock (Salimon et al. 2010):

- Higher lubricity leading to lower friction losses, yielding more power, and better fuel economy.
- Lower volatility resulting in decreased exhaust emissions.
- Higher viscosity indices.
- Higher shear stability.
- Higher detergency eliminating the need for detergent additives.
- Higher dispersancy.
- Rapid biodegradation and hence decreased environmental/toxicological hazards.

7.8 Development of Bio-lubricants Technical Properties

Vegetable plant oils are composed mostly of triacylglycerol (98 %) and contain of different fatty acids attached to a single molecule of glycerol. They also contain minor amounts of mono- and di-glycerols (0.5 %), free fatty acids (0.1 %), sterols (0.3 %), and tocopherols (0.1 %). Fatty acids are mainly long chain (C_{18}–C_{24}) unbranched aliphatic acids, with hydrogen atoms attached to carbons and other groups, and a carboxylic acid terminating the chain. The shortest non-ranched fatty acid chains (C_6) is water-soluble due to the presence of the polar –COOH group. By increasing the length of the chain, the fatty acid takes on oily or fatty characteristics and becomes increasingly less water-soluble. The carbon chain of a fully saturated fatty acid is straight. When hydrogen atoms are missing from adjacent carbon atoms, the carbons share a double bond instead of a single bond. This type of fatty acids are called un-saturated fatty acids. These acids have lower melting points than saturated fatty acids. The fatty acid is polyunsaturated if double bonds occur at multiple sites. Therefore fatty acids can be classified as being saturated, mono-, di-, tri-unsaturated, etc.

Mineral oils, on the other hand, are extremely complex mixtures of C_{20}–C_{50} hydrocarbons containing a range of linear alkanes (waxes), branched alkanes (paraffinics), alicyclic (naphthenic), olefinic, and aromatic species. They also contain significant amounts of heteroatoms, mainly sulfur. Mineral oils are more stable, cheaper, and more readily available than natural oils, and are also available in a wider range of viscosities. One issue regarding mineral oils is that oils derived from different fields have different characteristics. Another issue regarding mineral oils is the volatilization of low molecular weight components, which leads to a tendency to thicken during use. The presence of low molecular weight components also reduce the flash point of mineral oils compared to natural oils of the same viscosity.

Synthetic oil—Synthetic organic esters are another class of widely used lubricants. Examples include esters derived from C_8–C_{13} mono-alcohols and di-acids such as adipic acid (di-esters) and esters of C_5–C_{18} monoacids with neopentyl polyols, suchas pentaerythritol (PE, polyol esters). The presence of the ester group confers low temperature fluidity and reduces volatility at high temperatures. It also provides an affinity for metal surfaces. Esters were originally developed for the lubrication of aircraft jet engines but have subsequently found wide-spread use, particularly in applications where biodegradability is required.

There are some applications where performance requirements cannot be met by mineral oils, and it is necessary to chemically synthesize base lubricants with superior properties. There are a range of synthetic lubricants such as poly alpha-olefins (PAOs). These lubricants have characteristics similar to highly refined paraffinic mineral oils, but with a more narrowly defined molecular weight distribution. Alkyl benzenes are a class of synthetic hydrocarbons, although the lubricant industry now almost exclusively uses branched alkyl benzenes, as they have better low temperature fluidity. All synthetic lubricants are normally used as formulations containing the same types of functional additives as are used in mineral oils.

7.9 Bio-lubricant Limitation: Additives and Modification Process

It was found out that direct application of plant oils as lubricants are less favorable due to a variety of factors. Plant oils have poor oxidative and thermal stability, which is due to the presence of acyl groups. The presence of the glycerol backbone in oil gives tertiary b-hydrogen, which is thermally unstable. The chemical modification of plant oils by addition reactions to the double bonds constitutes a promising manner of obtaining valuable commercial products from renewable raw materials. In order to use plant oil-based lubricants with special additives, chemical modifications, de novo synthesis, breeding, and/or biotechnology play an important role. These methods improve the performance and stability of base oils in lubricating formulations. They also provide a sufficient capacity of plant-based oil substrates for green engineering.

As outlined in Chap. 1, lubricants and hydraulic fluids are materials that are composed of a base fluid and additives. In hydraulic fluids additives account for only 1 % of the formulation and 15 % of the total volume in motor lubricants. In transmission oils, additives constitute about 8 % of the formulation. Typically, additives are used as antioxidants, rust (corrosion) inhibitors, de-emulsifiers, wear reducers, PP depressors, and hydrolysis inhibitors. Most additives are common in mineral and plant oils, but the toxicity of currently used additives require research on the development and use of alternatives bio-based additives. Naturally occurring antioxidants such as tocopherol (vitamin E), L-ascorbic acid (vitamin C), esters of gallic acid, citric acid derivatives, or lipid-modified EDTA derivatives serve as synthetic metal scavengers and may be investigated as alternatives to the currently used toxic antioxidants. Chemical modifications such as epoxidation, *estolides* formation, and *tran* esterification of plant oils with polyols have been shown to improve the oxidative stability of plant oil based lubricants and to achieve optimal characteristics for extreme applications.

Plant oils and animal fats are increasingly used as green raw materials in various areas of industry. In the field of lubricants, environmental, and economic reasons lead to the utilization of plant oils and animal fats, or used oils and fats after appropriate chemical modifications. The temperature flow property of pant oils is extremely poor, and this limits their use at low operating temperatures, especially in automotive and industrial fluids. Plant oils have a tendency to form macro-crystalline structures at low temperatures through uniform stacking of the "bend" in the triacylglycerol backbone. Such macro-crystals restrict flow due to the loss of kinetic energy of individual molecules during self-stacking. Several di-ester compounds have been synthesized from commercially available oleic acid and common fatty acids.

The key steps in the three step synthesis of oleo-chemical di-esters includes **epoxidation and ring opening** of epoxidized oleic acid with different fatty acids (octanoic, nonanoic, lauric, myristic, palmitic, stearic, and behenic acids) using *p*-toluene sulfonic acid (PTSA) as a catalyst to yield mono-ester compounds. The esterification reaction of these compounds with butanol, iso-butanol, octanol, and 2-ethylhexanol was further carried out in the presence of 10 mol% H_2SO_4, producing the desired di-ester compounds. Not surprisingly, as the length of the mid-chain increase, a corresponding improvement in low temperature behavior is observed. This phenomenon is due to the increased ability of the long chain esters to disrupt macro crystalline formation at low temperatures. Another observation is the positive effect of branching at the chain end on the low temperature performance of the resultant products, which leads to the formation of micro crystalline structures rather than macro crystalline structures.

Estolides are a class of esters-based on plant oils and are synthesized by the formation of a **carbocation** at the site of unsaturation. This carbocation can undergo nucleophilic attack by other fatty acids, with or without carbocation migration along the length of the chain, to form an ester linkage. Estolides were developed to overcome some of the short-falls associated with plant oils, such as poor thermal oxidative stability and poor low temperature properties. Some deficiencies can be improved with the use of additives but usually at the expense of biodegradability, toxicity and cost.

Biodegradable organic polyesters derived from the transesterification/esterification of plant oils and branched neo polyols such as tri-methylol propan (TMP) and PE have been developed for various applications.

7.10 Bio-degradable Greases

Conventional lubricating greases based on mineral oils or on certain synthetic base oils are biodegradable only to a very limited extent. When they end up in the environment, such greases remain there for a long period of time, causing the threat of contamination to both fauna and flora. Lubricating greases mostly used are based on mineral oils with low biodegradability. In order to aid environmental protection, biodegradable lubricating greases have been developed based on lithium and calcium soaps and biodegradable synthetic esters and vegetable oils. These greases represent a suitable replacement for some conventional mineral greases. Unlike mineral-based greases when in contact with the environment, these greases biodegrade within a period ranging from several weeks to several months, depending on environmental conditions. Careful choice of base oils, thickeners, and additives, their optimal formulation ratios, and the use of modern technology, may produce lubricant greases which are not only biodegradable, but also have properties matching or even exceeding those of mineral-based conventional oils (Legisa et al. 2005).

7.11 Bio-lubricants Potential for Long-Term Use

Replacing hydrocarbon-based oils with biodegradable products is one of the ways to reduce adverse effects on the ecosystem caused by the use of lubricants. The use of low or no sulfur, low ash and phosphorous (low SAP) esters or polyglycol-based oils (intended for passenger car engine lubricants as substitutes for hydrocarbon-based oils) requires the preparation of a composition of lubricants with comparable tribological and functional properties. Thus there is need to develop biodegradable passenger car lubricants in combination with tribo-reactive materials, focusing on passenger car motor oils (PCMO) with reduced metal–organic additives. This being necessitated by the requirement to reduce the ash build-up in the treatment system and to improve its efficiency and lifetime. High fuel efficiency and long drain intervals are also necessary.

Additionally, these oils have to be biodegradable and non-toxic to aqueous environments, in line with other international standards. In a modern diesel or gasoline engine, the engine oils has to fulfill a number of functions, such as lubricating and cooling the system, protecting against wear, handling of soot and particles with low deposit tendency etc. The study of the biodegradability, toxicity, and tribological properties has been carried out for newly developed prototype engine bio-oils. Furthermore, plasma sprayed tribo-reactive coatings have been deposited on cast iron piston rings and studied for their tribological properties. Finally, the behaviors of bio-oils and plasma-sprayed tribo-reactive coatings on piston rings have been screened in a real engine.

A tremendous demand for plant oils in the lubricant industry is expected over the next few years because plant oils are natural, renewable, non-toxic, non-polluting, and cheaper than synthetic oils. They will become an important class of base stocks for lubricant formulations due to their positive qualities. Due to growing environmental concerns, plant oils are finding their way into lubricants for industrial and transportation purposes. Plant oils, in comparison to mineral oils have different properties due to their unique chemical structures. Bio oils have better lubrication ability, viscosity indices, and superior anticorrosion properties, which are due to the higher affinity of plant oils to metal surfaces. In addition, flash points are greater than 300 °C classify plant oils as non-flammable liquids. To improvement characteristics such as sensitivity to hydrolysis and oxidative attacks, poor low temperature behavior, and low viscosity index coefficients, plant oils may be chemically modified. Renewable-based lubricants are being considered as potential alternatives to petroleum-based lubricants for various reasons, mainly increased environmental sensitivity. However, understanding the tribological performance of such vegetable-based lubricants under elevated temperatures is critical for their industrial implementation.

Finally, let it be noted that in practical applications, lubricants may accumulate certain products from ageing and wear, making used oils often considerably more hazardous for the environment than fresh ones. The same principle can also apply to bio-lubricants which are less harmful prior to their use (Legisa et al. 2005).

7.12 Biodegradation Accelerants for Lubricants

One of the main advantages of petroleum bases over vegetable oils is their higher versatility, which involves producing bases with higher viscosities and lower pour points while maintaining high levels of stability. The production of fatty acid or triacylglycerol estolides allows for the production of biodegradable bases with improved properties and opens a promising field for the oleo-chemistry applied to bio-lubricant production (Willing 2001).

Can the biodegradability of unreadily biodegradable lubricants like mineral petroleum and PAOs be improved by formulating so-called 'biodegradation accelerants'? Sure this can be done, as it is known that biodegradability is not an exact property or characteristic of a substance, but a 'system' concept, i.e. the conditions within an entire system determines whether or not a substance within it is biodegraded. Studies have shown that many phosphorous and nitrogenous compounds are highly effective in promoting hydrocarbons to biodegrade and have been successfully employed in the remediation of petroleum polluted areas such as water and soil.

Biodegradability of unreadily biodegradable lubricating base oils such as mineral oil and high molecular weight polyalphaolefin can be greatly promoted by phosphorous and nitrogenous accelerants, due to the activation of microbial enzyme and microbial growth (Boshui et al. 2008). Thus this breakthrough can be used to promote biodegradability in other lubricants other than the vegetable based lubricants for environmental reasons.

7.13 Bio-based Lubricants Market and Potential

There have been some inroads to consider adoption of bio-preferred products, in environmentally sensitive application areas to replace mineral oils, provided the bio-preferred products must compete on performance and price. The major sources of the vegetable based lubricant base have been soya-beans and rapeseed.

The soyabean base oil has achieved the most success is transformer dielectric fluid, also referred to as transformer insulating oil. The acceptability has been high as the product shows both a performance and an economic advantage over mineral oil products in terms of a much increased fire point, increased service life of the transformer due to extended life of the insulating paper and the potential for much lower cost spill remediation due to favorable biodegradability and lower toxicity characteristics.

The hydraulic fluid area has grown but the growth rate has been slow and still only represents a small percentage of the overall market for these fluids. The segment that has achieved significant success is elevator oils. Bio-based hydraulic fluids are finding increased use at military bases, national laboratories and national parks. Metalworking fluids have been a successful application for soyabean oil lubricants, and a rolling oil lubricant has been developed and is being used commercially today in plants around the world (Bremmer and Plonsker 2008).

The total-loss lubricant area has reportedly not progressed very far, as bar and chain oils have not seen very much growth, although bio-lubricant products are available to consumers for this application. Wire rope oils had the distinction of being one of the few areas where there was a regulation that promoted the use of biodegradable oils. It is unclear to what degree this is still the case today. In any case, the use of bio-lubricants in this application is small and does not seem to have grown.

The largest single potential market for lubricants, crankcase oils, has not yet developed a product qualified to meet industry standards. Work still goes on in this area due to the potential cost advantage opportunity and products are close to being qualified although the bio oil only represents a small percentage at present of the total formulated oil product. However, a small percentage in this market still represents a multi-million barrel potential if broadly adopted. The fact remains that the bio-lubricants are more expensive and, without regulations, the choice is usually for the cheaper mineral oil based product. Volume growth is still small although revenue growth is larger because of the higher price of the bio-lubes (Bremmer and Plonsker 2008).

7.14 Conclusion

In conclusion, lubricants are an important part of new strategies, policies, and subsidies, which aid in the reduction of the dependence on mineral oil and other nonrenewable sources. Products with toxicological and ecological issues must be excluded from further use in lubricants, if they pose a significant health risk. Plant oils may be used in almost all automotive and industrial applications. As much as it is difficult to find a balance between the economic possibilities of bio-lubricants and their ecological requirements, researchers should keenly work on improving current findings in this regard. It was also noted that marketing of these more tolerable environmental impact lubricants is proceeding more slowly than expected, because legal regulations have not yet made it mandatory for lubricant consumers to use less harmful lubricants, which are currently more expensive than conventional lubricants.

Revision Questions

1. *List advantages of bio-lubricants over mineral oils?*
2. *What key modifications are in improving bio-lubricants to effectively lubricate?*
3. *What are the major current short comings of bio lubricants?*
4. *Name one key additive for the bio oils?*
5. *What danger is there in putting more and more additives in bio oils?*
6. *What do understand by bio-degradability accelerants, and where are they used?*

References

Boshui C, Jianhua F, Ling D, Xia S, Jiu W (2008) Enhancement of biodegradability of lubricants by biodegradation accelerators, Lubr Sci 20:311–317, Wiley InterScience (www.interscience.wiley.com) doi:10.1002/ls.68

Bremmer BJ, Plonsker L (2008) Bio-based lubricants: a market opportunity study update. United Soyabean Board, Omni Tech International Ltd

Chauhan PS, Chhibber VK (2013) Non-edible oils as potential source for bio lubricant production and future prospects in india: a review. Indian J Appl Res Chem 3(5)

Jaina AK, Suhane A (2013) Capability of bio-lubricants as alternative lubricant in industrial and maintenance applications. Int J Curr Eng Technol 3, International Press Corporation (INPRESSCO), http://inpressco.com/category/ijcet. Mar 2013

Legisa I, Picek M, Nahal K (2005) Some Experience with biodegradable lubricants. J Synth Lubr 13(4)

Salimon J, Salih N, Yousif E (2010) Bio-lubricants: raw materials, chemical modifications and environmental benefits. Eur J Lipid Sci Technol 112:519–530, WILEY, Verlarge, Weinhem

Willing A (2001) Lubricants based on renewable resources—an environmentally compatible alternative to mineral oil products. Chemosphere 43:89–98, Elsevier Science Ltd

Chapter 8
Lubricant Life Cycle Assessment

Abstract A general life cycle assessment (LCA) study of base oils used in the manufacture of lubricating fluid is given in this chapter. The scenarios of mineral oil, a synthetic ester, and a vegetable (rapeseed triglyceride oil) are investigated. The review of LCA leads to drawing conclusions concerning the application of LCA models as evaluation tools for the development of environmentally adapted lubricants in line with IS0 14000-type industrial standard.

8.1 Introduction

Up until the 1980s the problems facing lubricant formulators and marketers were simply those of meeting the technical requirements set by Original Equipment Manufactures (OEMs) and customers. However in the last 10 years a whole extra set of requirements have been introduced, through legislation, public concern and OEM pressures, dealing with a lubricant's impact on the natural environment. Such impacts must be examined through a lubricants full life cycle all the way from the original product development program through to ultimate disposal. Although a full life cycle analysis is needed in order to assess the total impact of lubricants in the environment, effort is made to highlight some recent developments in the following critical areas (Vlg et al. 2002):

- Choices during formulation, especially those dictated by current and emerging legislation,
- Impacts during use of the product including such aspects as fuel economy, oil consumption and tailpipe emissions,
- Fate in the environment, with emphasis on the disposal and recycling of used lubricants.

In the quest to develop an environmentally adapted lubricants (EALs), a number steps may be considered towards more ecological ends, whereas for an overview of environmental impact, life cycle assessment (LCA) is a useful tool. In the field

© Springer International Publishing Switzerland 2016   103
I. Madanhire and C. Mbohwa, *Mitigating Environmental Impact of Petroleum Lubricants*, DOI 10.1007/978-3-319-31358-0_8

of lubricants, a variety of stages can be improved. LCA can be used as a basis for discussion of the most important stages during the development process. Although LCA is not new, it is not commonly used in the development of lubricants. Some progress has been made towards understanding the complex life of a lubricant from 'cradle to grave' including the environmental and socio-economic impact of replacing mineral oil with rapeseed oil as a base oil in the manufacture of hydraulic fluid. They found that the 'global warming potential' was considerably lower for rapeseed oil than for mineral oil. Nevertheless, more data are required for LCA methodology to be used when developing future environmentally adapted lubricants (Vlg et al. 2002). The three types of bases which entail: mineral oil, rapeseed oil and a synthetic ester would be reviewed for mobile hydraulic systems in forestry harvesting operations, because of the interest in this sector in replacing mineral-oil based hydraulic fluids.

The differences in the physical and chemical properties of these base fluid groups have a profound influence on the properties of the finished lubricants. For instance, the technically useful life of vegetable oil based hydraulic fluids is generally shorter than that of other fluid types, but the degree of renewability and biodegradability are higher.

8.2 Petroleum Mineral Base Oil

Mineral base oils are produced by vacuum distillation of crude oil. In this process, the residue from atmospheric distillation of crude oil is further distilled under vacuum conditions to produce a range of vacuum distillate fractions. Solvent extraction and/or hydro-finishing is then used to increase the viscosity index, enhance the color, and convert undesirable chemical structures (such as unsaturated and aromatic hydrocarbons) to less chemically reactive species. Finally, solvent de-waxing is used to reduce the wax content of the base oils and to improve the low-temperature performance of the lubricant.

8.3 Synthetic Ester Base Oil

The synthetic esters are composed partly of renewable material and partly of synthetic material of petrochemical origin. The most common ester types comprise fatty acids of vegetable or animal origin, and a synthetic poly-alcohol (polyol) made from petrochemical feed stocks. The production of the ester usually involves a chemical process known as trans-esterification. In this process, a natural triglyceride oil of vegetable or animal origin is converted into an ester by the exchange of the alcohol part of the ester molecule. The natural glycerine polyol present in the vegetable oil is replaced by the petrochemical polyol, and the glycerine is removed. The trans-esterification process involves the use of heat, a catalyst, and an excess of the synthetic polyol. A large variety of ester structures, with a range of properties, is thus accessible through this methodology, since many different natural esters (containing a variety of different fatty acids), and several types of synthetic polyol are available. It is also possible to perform a straightforward esterification using free fatty acids and the selected polyol. This procedure is used if a suitable source of a pure fatty acid is available, or if very special properties are desired.

8.4 Vegetable Base Oils

Lubricants derived from vegetable and animal sources are used in many applications, e.g., automotive oils, spindle lubrication, transmission fluids, metal-cutting fluids, and gear lubricants. The main advantages of vegetable oils are that they are readily available, have a lower price than synthetic esters, are 100 % renewable, and are readily biodegradable. The technological drawbacks are their sensitivity to high temperatures and poorer low-temperature performance and oxidative stability.

8.5 LCA for Lubricants

LCA is one of several tools for environmental management (e.g., risk assessment, environmental impact assessment, and environmental auditing). It identifies and quantifies the energy and materials used from 'cradle to grave' to establish the consumption of energy, material used, and emissions. Normally, LCA does not include the economic and social aspects of a product. Since all tools for environmental management have their limitations, it is of importance to understand these in a specific LCA study. LCA methodology describes the manufacture and use of lubricants to increase the understanding of different parts of the process in a systematic way. Lubricant companies that decide to compete in the EAL market should have to consider total product stewardship, requiring the 'cradle to grave' approach to be used in product design. Consideration must therefore be given to aspects such as the origin of the raw material, and the production, packaging, transportation, application, and disposal of the product.

All finished hydraulic fluids considered contain performance additives, but the relative types and amounts are similar. The data utilized for the impact of cultivation, pressing, and extraction of rapeseed oil.

8.6 Mineral Base Lubricant LCA

The life cycle of a hydraulic fluid made from mineral oil is shown schematically in Fig. 8.1. Distillation is the key operation in petroleum refining. Base oils for lubricants have very high boiling ranges, and are usually produced by a vacuum distillation step following two-stage distillation. Long-chain alkanes of high boiling

Fig. 8.1 Life cycle of
mineral-based hydraulic fluid
in a total loss application

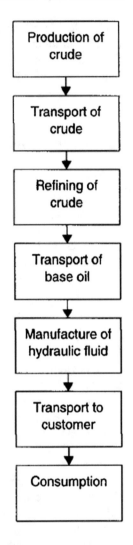

points have most of the properties that are desirable for the formulation of lubricant oils. Middle East crude oils are a frequent source of lubricant base oils.

When the crude oil has been refined it is shipped to the lubricant blending plant. During the blending process, energy in the form of steam and electricity is used. The finished products are eventually filled in various kinds of packages before sale, but the impact of the packaging has been excluded from this LCA. The product is then transported by road to the customer. When the product has reached its destination, it is used in a harvester. During the biodegradation of any spilled hydraulic fluid, carbon dioxide (CO_2), a potent greenhouse gas (GHG), is formed and released into the atmosphere (Fitch and Gebarin 2003). The eventual breakdown and biodegradation of the mineral base oil contribute to the GWP calculation, in the form of CO_2, released, since this CO_2, is not a part of the natural carbon cycle.

8.7 Synthetic Ester Lubricant LCA

The process steps used in this LCA to describe the manufacture of a synthetic ester-based hydraulic fluid were shown in Fig. 8.2. The synthetic ester (Fig. 8.3) used in this study was produced by the trans-esterification method from rapeseed oil (a natural triglyceride ester) and a petroleum-based polyol. Typical

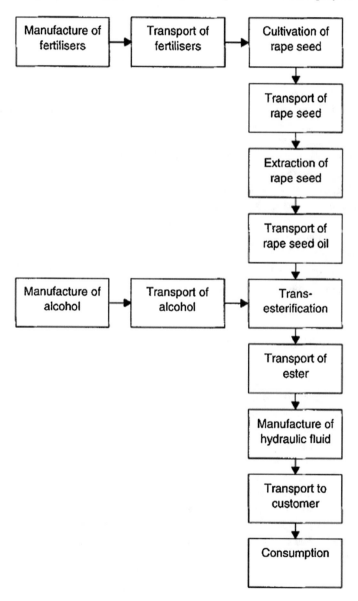

Fig. 8.2 Material flows for the production of a synthetic ester-based hydraulic fluid

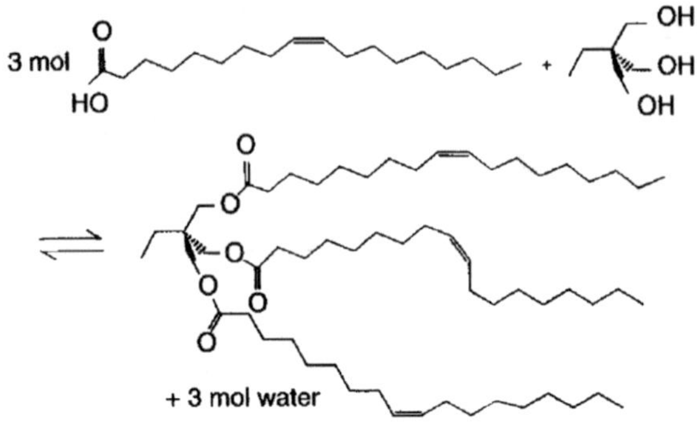

Fig. 8.3 Formation of
a typical synthetic ester,
trimethylol propane trioleate
(TMP trioleate)

petroleum-based polyols used in the esterification process are trimethylol propane, pentaerythritol, and neopentyl glycol. In the esterification process the rapeseed oil is heated up, and the catalytic agent and the polyol are added. Glycerol is separated and discharged. The excess polyol is removed by evaporation. The exact values of the energy and resources consumed in the esterification process are not available, but data from a similar product have been used.

The blending process for synthetic ester-based hydraulic fluids is identical to the process described above, and the transportation demand is also the same. However, in the calculation of global warming potential none of the CO_2, formed during the biodegradation of the rapeseed oil part is included, since CO_2, from this source is a part of the natural carbon cycle. This, of course, constitutes one of the major advantages of using renewable resources in the production of lubricant (or other goods). This fact is thus expressed in the output of the LCA.

8.8 Rapeseed Base Oil LCA

Rapeseed oil and meal cake products are used throughout the food and feed industry, and are also important in various applications such as lubricants and the bio-fuel rapeseed methyl ester. The material flow for the production of rapeseed oil is shown in Fig. 8.4. Production of rapeseed oil is achieved by pressing the seeds. The remainder is then extracted with hexane, a petrochemical solvent. Following the extraction step, the meal cake is treated with steam to remove any remaining hexane solvent. The blending process for rapeseed oil-based hydraulic fluids is identical to the process described for mineral-based hydraulic fluids, and the transportation demand is also the same. However, in the calculation of global warming potential none of the CO_2, formed during the biodegradation of the rapeseed oil is included, since CO_2, from this source is a part of the natural carbon cycle.

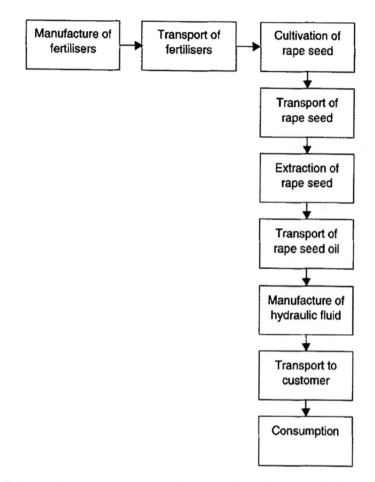

Fig. 8.4 Material flows for the production of rapeseed oil-based hydraulic fluid in a total loss application

8.9 Environmental Impact

It is difficult to express environmental impact as a single scalar number. Therefore, in accordance with the LCA methodology, the environmental impact indicators, acidification potential and global warming potential, were used. The emissions used to estimate global warming potential are CO_2, methane (CH_4), and dinitrous oxide (N_2O). Carbon dioxide is considered to be the most significant greenhouse gas due to the large amounts discharged. However, when analyzing agricultural and forest systems, discharge of CH_4 and N_2O are of importance since these gases have a larger effect per mole discharged. The weighting factors used in this study are given in Table 8.1.

Table 8.1 Weighting factors used to calculate the global warming potential (time horizon 20 years)

Emission	Weighting
CO_2	1
CH_4	62
N_2O	290

Table 8.2 Weighting factors for the acidification potential

Emission	Weighting
SO_2	1
NO_x	0.7
HCl	0.88
NH_3	1.88

Acidification is caused by emissions of gases such as sulphur oxide (SO_2), hydrochloric acid (HCl), nitrous oxides (NO_x), and ammonia (NH_3), which, combined with water, oxygen, and ozone in the atmosphere, cause acid rain. Acid rain causes damage to soil and plants directly or indirectly by oxidation in the soil. Acidification is measured as SO_2 equivalents. The acidification potential of the production of the base oils studied included contributions from SO_2, HCl, NO_x and NH_3. The weighting factors are given in Table 8.2.

The LCAs show that the rapeseed oil-based hydraulic fluid has the lowest environmental effect with regard to both the global warming and the acidification potential. The energy used to produce the rapeseed oil-based hydraulic fluid is also the lowest. The global warming potential for the mineral-based hydraulic fluid is three times higher than for the rapeseed-based one. The synthetic ester-based hydraulic fluid is between the two, with almost double the potential of the rapeseed oil (Fig. 8.5).It is evident that the potential contributions of CO_2 to global warming potential from the manufacturing of mineral-based oils are largely related to the use of fossil fuel during the refining process (Fig. 8.6). The CO_2 contributions are not as important for the rapeseed oil or the synthetic ester. However, the contribution of N_2O to global warming potential is 57 and 29 % for the rapeseed and the synthetic ester, respectively. The high contribution of N_2O originates from the use of fertilizers during the cultivation of the rapeseed.

In order to relate the global warming potential values calculated for these hydraulic fluid types to some external factor, hypothetical calculations on fuel saving were performed. It has been argued that it would be possible to save approximately 3 % of the fuel consumption of the harvester by using the hydraulic fluid giving the best performance (Fig. 8.7).

It should be noted that the global warming potential in the fuel-saving case is only measured as CO_2 emissions. By using the optimal lubricant solution for all

Fig. 8.5 The global warming potential contribution for the base fluids

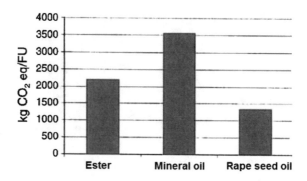

Fig. 8.6 The relative contribution to global warming potential of CO_2, CH_4 and N_2O for the base fluids

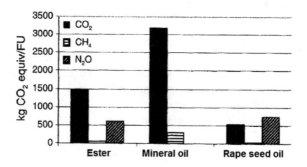

Fig. 8.7 Global warming potential saved when 3 % less diesel is used for the base fluids

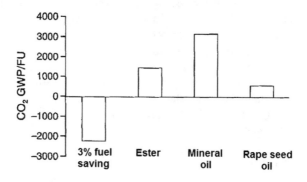

energy-consuming systems of the harvester, total fuel savings of 7–8 % should be possible.

The total acidification potential and that of the base fluids reported appear in Fig. 8.8. The energy consumption per functional unit during the production of a mineral base fluid is greater than for a synthetic ester or a rapeseed oil.

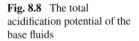

Fig. 8.8 The total
acidification potential of the
base fluids

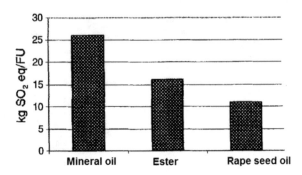

8.10 Conclusion

The LCAs show that the production of vegetable oil has the lowest global warming potential, energy consumption, and acidification potential. The life of the
hydraulic fluid is not considered to be of importance in total loss application,
where fresh fluid is frequently added to the system. In other applications, where
the leakage rate is much lower, and the residence times could be substantially
increased, it is reasonable to expect that a difference in useful product life would
influence the outcome of the LCAs.

An environmentally adapted lubricant as that which is readily biodegradable in
one of several internationally recognized test methods, or possessing low toxicity towards aquatic life, or both. When reasoning about sustainable development
each part of society has to assume responsibility for the consequences of its own
actions. The difference between the environmental impact rating obtained for the
synthetic ester and that for the rapeseed oil can be explained by the petroleum-
based polyol in itself contributing negatively to the environmental impact scoring.
By switching to a polyol with a less negative impact, an improved scoring would
be obtained for a synthetic ester. Thus, it is desirable to utilize this type of LCA as
a design tool in the quest for improved formulations for environmentally adapted
lubricants.

Revision Questions

1. *What is an EAL?*
2. *Highlight major differences among mineral oil, synthetic oil and vegetable oil?*
3. *What causes major polluting effect from the mineral oil LCA?*
4. *Which gases are responsible for acidification and global warming?*
5. *What other advantage is inherent in the vegetable lubricants?*

References

Fitch JC, Gebarin S (2003) Best practices in bulk lubricant storage and handling. Noria Corporation, Tulsa (OK 74105 USA)

Vlg C, Marby A, Kopp M, Furberg L, Norrby T (2002) A comparative life cycle assessment of the manufacture of base fluids for lubricants. J Synth Lubr 19(1)

Chapter 9
Environment and the Economics of Long Drain Intervals

Abstract Recent technological developments have revolutionized the lubricant industry by spurring a widespread quality improvement in both base oils and additives, giving superior finished lubricants. The upgrading of base oils is being brought about by more demanding requirements from original equipment manufacturers (OEMs), by government regulations, and through consumer awareness, environmental concerns, decreasing supply of high-quality lubricant-bearing crudes, and expanding markets worldwide. Present-day lubricant demand is for maximum oxidation stability, superior low-temperature performance, low volatility, and improved additive response, which are difficult to achieve through conventional processing. All this in a nutshell seek to satisfy environmental requirements and extended drain economics. Hence a serious consideration is being muted to introduce a regulation to enforce bio-degradable lubricants which are vegetable based. Condition monitoring and proactive maintenance are critical tools for achieving significant improvement in tribological performance of mechanical components and extended lubricant life.

9.1 Introduction

Technically, base oil manufacturers are mindful of the need to make base oils using developing technologies, to produce high-performance base oils economically, maximize research in the areas of lubricant processing and of additive development. Original equipment manufacturers (OEMs) and lubricant formulators are seeking higher-performance and longer-life base oils, as well as higher specifications with respect to quality requirements. In view of the above, along with environmental considerations, base oil manufacturers are forced to produce improved base oils by implementing newer technologies. 'New-generation' base oils exceed the performance of conventional lubricants in terms of volatility, oxidative stability, low carbon-forming tendency, and additive response. The key to such base oils is the conversion process, which chemically

© Springer International Publishing Switzerland 2016
I. Madanhire and C. Mbohwa, *Mitigating Environmental Impact of Petroleum Lubricants*, DOI 10.1007/978-3-319-31358-0_9

transforms unwanted reactive species into the desired lubricant components, across a wide range of crude oils. This chapter reviews the technological options for increasing the quality of lubricants being produced (Anwar et al. 2002).

Another scenario involves the physical location of operating equipment in sensitive environmental areas. This may severely limit the choice of traditional lubricants and lubrication techniques as the ecological properties of candidate fluids take precedent over lubricant performance properties. Simultaneously and conversely, increased demands on the chemical and petroleum industries to engineer materials that are environmentally safe and friendly inherently limit the application range of these products. One must realize that demands, regulations and obligations about the environmental, ecological, and disposal issues must be met and dealt with while continuing to produce competitive products (Pinchuk et al. 2010).

9.2 Lubricant Consumption Control

In addition to significant contributions to fuel economy, conservation of petroleum resources with attendant reduced contribution to emissions is achieved by enhancing lubricant economy in service as given by Table 9.1 over the years.

Modern, low emission engine designs create significant lubrication challenges for additive chemistry to control deposits and wear, and maintain long-term low emissions performance. The combination of reduced oil consumption and extended drain intervals greatly increases the load on the lubricant (Fig. 9.1). Lubricant additives assist in reducing oil consumption by modifying the physical properties of the lubricant such as viscosity, and enable the use of less volatile base fluids. Additive components also help control wear and deposits, maintaining low oil consumption throughout a vehicle's life (ATC 2007). Thus lubricant additive development has permitted extended oil drain service intervals, enhanced fuel

Table 9.1 Engine oil use trends

Vehicle engineering and lubricant additive development				
Higher performance, with reduced lubricant quantity for gasoline passenger cars				
(data based on top specification engines)				
Model year	1949	1972	1992	2005
Power (kW)	25	74	96	120
Power density (kW/L)	21	37	45	60
Oil fill (L)	3.0	3.7	3.5	3.5
Oil consumption (L/1000 km)	0.5	0.25	0.1	0.1
Oil change interval (km)	1500	5000	15,000	30,000
Oil flush at oil change	Yes	No	No	No
Total oil used after 30,000 km (L)	87.0	29.8	10.0	6.5
Average fuel consumption (L/100 km)	12	10	7	7
Engine durability (1000 km)	<100	175	250	250

Fig. 9.1 Stress on passenger car lubricant

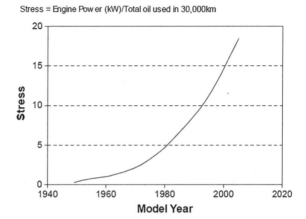

efficiency and improved engine durability. These factors have contributed measurably to conserving petroleum and other natural resources.

9.3 Extended Drain Interval

In extreme environment applications, fluid degradation or contamination is usually premature. This alone can result in untimely machinery breakdown as undetected additive depletion and contamination will cause catastrophic failure (Fig. 9.2). In most cases, the direct cost of failure of a flawed fluid management program far outweighs the indirect 'cost savings' of a successful one.

In Fig. 9.2 it is shown how a typical lubricant degradation occurs over time. In this case, an ISO 46 (orange line) anti-wear hydraulic fluid was operating at 150 °C. Initial antioxidant and anti-wear levels were at 100 % in the 'new' oil (red and blue lines respectively). The originating TAN (Total Acid Number) was 0.2 KOH/mg (green line). As the fluid is stressed in operation a depletion in antioxidant after only 30 h of operation causes a significant increase in AN therefore increasing the fluid viscosity and resulting in premature lubricant failure.

Metallurgical alloys used in bearings, and gears, etc., are assembled and tested at specific operating speeds and temperatures. At elevated speeds and temperatures, these alloys will deform and expand in unpredictable and non-reproducible dimensions. These deformations will normally reduce clearances allowed for lubrication and change spherical rotation to 'erratic'. This in itself starves the lubrication points, generating more heat, more deformation, increased vibration, and finally failure. Ball bearings exposed to heat levels beyond their original design and they lose their spherical form, and created scuff marks on the surface by due to sliding (Pinchuk et al. 2010).

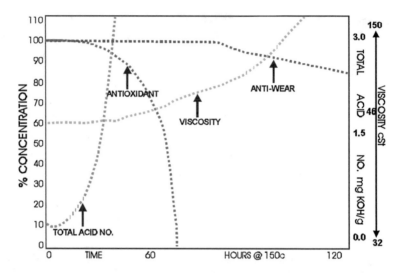

Fig. 9.2 Typical lubricant degradation pattern

For machinery operating in extreme environments, in addition to wide fluctuations in temperature, tough environmental laws render lubricant disposal very costly. Fluid handling in and out of a subterranean mine could also cumbersome and a major safety consideration. Fluid degradation as depicted in Fig. 9.2 is inevitable. To counter this deterioration, an aggressive lubricant sampling program is established and degradation curve tracked for each piece of equipment. Minimum additive levels were determined and were dosed back into the lubricant in situ as needed (Fig. 9.3).

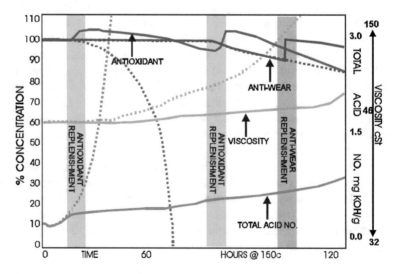

Fig. 9.3 Lubricant degradation with additive dose replenishment

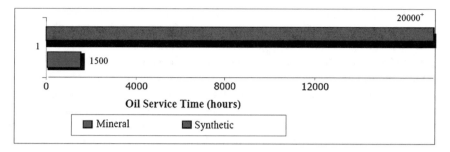

Fig. 9.4 Hydraulic oil—synthetic and mineral oil comparison

If the same typical fluid degradation graph as Fig. 9.2 except that the antioxidant was replenished at two different intervals and the anti-wear was replenished once. As a result of the first antioxidant replenishment the TAN stabilized and therefore kept the viscosity from increasing. This was repeated a second time as the antioxidant reached its minimal level again. The same was done for the anti-wear additive as its level reached minimum tolerance.

In Fig. 9.4 overall service life of a hydraulic synthetic fluid is compared to its original mineral-based counterpart.

Lubricants operating in extreme environments require a collaborative conditioning monitoring program to be effective. Technological advances in metallurgy, lubricants, and composites are still in their embryonic stage. Advancements in one field pose new challenges to the others. Communication and collaboration is mandatory in these applications as there is an inherent dependency on all contributing aspects for extended drain to get the best out of lubricant in use as well as reducing waste used oil thrown into the environment.

9.4 Technology to Enhance Bio-lubricants for Extended Drain

While bio-lubricants have been found to be a good substitute for mineral base oils their instability is their great short coming hence need for relevant technology to address that. The oxidative instability of base oil sources such as soybean oil or rapeseed oil, and other vegetable oils is due to the presence of polyunsaturated fatty acids, as in linoleic acid and linolenic acid. Efforts have been made to modify the soybean oil to moderate the effects of these materials to provide a more stable material and a product more competitive in performance to mineral oil-based lubricants. The different method have been found to do this.

Additive technology: Oil formulations, whether they are bio-based or mineral oil, are generally regarded as proprietary information. It is difficult to know the identity of the additives added to the vegetable oil in those products that have been developed. Patents do not always give the answers. There are some indications from the patent literature that some investigators have developed stable products based on additive technology. An interesting study on the interaction of bio-oils with EP additives has been published elsewhere in the book.

Chemical transformation and polymerization: There are many different ways to modify the multifunctional vegetable oils. Some reported changes that address the polyunsaturated problem include alkylation, acylation, hydro-formylation, hydrogenation, oligomerization (polymerization) and epoxidation. No commercially available products based on these modifications have been identified, except perhaps on polymerization. There is some indication that polymerization is used to produce a stable hydraulic oil for elevator applications. Although it is possible that the end product oils are acceptably stable in the intended applications for some of these reported transformations, the added processing cost is likely to be a hurdle limiting broad commercial adoption, except perhaps in the case of polymerization.

Transesterification: This chemical transformation fits in with the above category but is isolated here because it is likely used in some bio-lubricant products. It is well known that improved lubricant performance can be achieved by replacing the glycerol part of the vegetable oil with other polyols, such a tri-methylolpropane, neopentyl glycol or pentaerythritol. These materials are considered to be semi-synthetics and are biodegradable but not entirely renewable.

Genetic modification: Research on genetic modification of soyabean oil and other vegetable oils has been an ongoing effort for many years. Much of this research has focused on reducing both saturated and polyunsaturated fatty acids and some varieties are already being marketed with improved cold flow properties and somewhat increased oxidative stability. These advancements are significant for industrial lubricants and samples are being evaluated by Valvoline and others for use as high temperature engine oils and hydraulic fluids. Also there are investigations for room to elevate the total oil content above the current average of 11.7 lb per bushel of beans. Plans have already been announced for the commercialization and sales of high oleic soyabean oil (HOSO). This development would greatly reduce the chemical entities that generate the stability problems, linoleic and linolenic acids. These reductions are made up by the substantial increase in the oleic acid component. It is obvious that this new soybean oil will have a substantial increase in stability and performance over the unmodified material. The future availability of this product should provide a boost for the bio-lubricant industry. Certainly, formulations will have to be modified to take advantage of the new soybean oil composition. In fact, this may provide an opening for those companies that have not been able to produce lubricants that meet the specifications required. This could be particularly true in the crankcase area where stability is a major problem.

It is believed that this chemistry could hold great promise for future modified soybean oils. One unresolved question is of the biodegradability of the new potential products and whether they would they be considered renewable or synthetic. All of the above approaches are aimed at improving the stability and performance of soybean oil and come at some additional cost beyond the feedstock oil itself. This is part of the reason for the higher price for the successful soybean oil products when compared to the mineral oil competition. Although mineral oils have seen dramatic price variations in recent years, so have vegetable oils. In some cases, the differences between the two product bases have narrowed considerably.

9.5 Bio-lubricant Base Oil Market

Crankcase oils are by far the largest segment of the global lubricants market, representing a demand in excess of one billion gallons per year in the US alone. It is also a segment which is extremely demanding in performance requirements, particularly in long-term oxidative stability where vegetable oils face inherent barriers not yet successfully surmounted. While the cost and technical sophistication required for research and qualification to industry standards is forbidding, the longer term opportunity has drawn the attention of the major industry leaders. A number of companies are competing to put engine oils on the market. Such a perfect example is engine oil for high mileage cars for a new and profitable market segment. They saw the use of a renewable component as a way to come up with yet another unique product.

The extremely high viscosity index and very low volatility of soybean oil (and other bio-fluids) Attracted interest because of the potential to allow savings in viscosity modifier treatment to obtain a given viscosity grade. In contrast, the need for additives to stabilize soybean oil at high and low temperatures was recognized. The cost of finished engine oil containing a bio-fluid and extra additives needed to be balanced with market profit opportunities. With changing soybean costs relative to petroleum base oil costs during early stages of the development program and currently with the rapid changes in costs of all raw materials, economic considerations on any given day were recognized as uncertain at best. Reputable marketers of high quality engine oil take all steps necessary to meet performance standards set cooperatively by industry participants (oil marketers, additive suppliers, engine builders, etc.).

For mineral oils and poly alpha olefins, API guidelines allow engine test results on some oils to be used as appropriate in the evaluation of others. However, API recently reaffirmed that all required Sequence Engine Tests must be run and passed with oil containing any bio-fluid content. A reduction in soybean oil content from 20 to 5 %, and further modification of antioxidants, improved Sequence IIIG performance, but a test pass was not yet obtained. At this point, Valvoline efforts shifted to engine oil with 5 % high oleic soybean oil. Performance extremely close to a passing Sequence IIIG has been obtained. A new bench test, the ROBO, has been used to predict Sequence IIIG performance and to select the best candidate for testing. Other engine test strategies carry forward from mid-oleic testing to the high oleic soybean oil.

If formulators simply replace mineral oil base stock in a lubricant with a bio-fluid component, even at low levels, performance of the lubricant will likely be degraded. Additive selection can regain performance lost by addition of a bio-fluid. The lubricant containing a bio-fluid component must be fully tested to prevent poor performance in customer equipment. A reputable oil marketer is needed to conduct the necessary testing in the laboratory and in field tests to ensure performance in customer equipment.

9.6 Transformer Insulating Fluids

Transformer dielectric fluid has undergone various significant changes over recent decades. At one time, polychlorinated biphenyls (PCBs) were used extensively. The fire retardant nature of this chemical was thought to be a highly desirable property. The disadvantage was its toxic nature and the lack of biodegradability. Any spills of the product dictated an extensive and expensive clean-up process to remove all traces of the material from the soil. In the 1970s, PCBs were banned due to their health and environmental hazards. New transformer oils were developed. They included naphthenic mineral oil which was the major product in early 2000 when the total U.S. consumption was about 45 million gallons per year. More specialty type products were also developed and marketed including high molecular weight hydrocarbons, synthetic esters and silicone fluid. The fluid was proven to enhance transformer performance, was environmentally friendly and could be used in a variety of electrical applications.

The present generation of vegetable-oil-based transformer fluids may be characterized as Premium products offering performance which in some respects is far superior to the conventional mineral-oil-based products they replace. In addition to biodegradability and low toxicity with corresponding spill remediation savings potential, they offer the safety advantages of a fire point above 300 °C versus 145 °C for the mineral oil product. The service life of a transformer is substantially increased due to the typically five times greater life of the insulating papers inserted between the windings. In contrast, the present generation of bio-based fluids are not as tolerant of extremely low temperatures as are the mineral oil products, typically limiting installations at present to exclude areas subject to extreme winter conditions. In addition, the excellent performance experienced in the US market is linked to the standard use here of sealed transformer systems; in Europe, the standard vented systems in use place a higher demand on oxidative stability of the fluid used.

9.7 Elevator Hydraulic Fluid

Although the commercial market has been slow to adopt of this fluid, the product has the advantage of not requiring heating in the winter and cooling in the summer because of its high viscosity index. Their primary customers are universities (Penn State is a large consumer) and hospitals. The price for these oils is about $15.00/gallon in drums compared to $9.50/gal for the corresponding mineral oil product. This price difference provides a competitive hurdle today except where considerable weight is placed on environmental considerations.

There is substantial market potential and this should attract the attention of the bio-lubricant business. The realistic hurdle in this market, like many others, is that there are no regulations to mandate adoption when the price is substantially higher for the bio-based materials. Other strides made so far in these products include stability in viscosity over a range of temperatures, fire resistance, sustainability and biodegradability. The cost of a spill remediation using a mineral oil product could be an offset to the higher price of the base oil, although at present this benefit is subject to varying regional regulatory guidelines.

9.8 Other Hydraulic Fluids

After crankcase oil, the hydraulic fluid area represents the largest volume potential. Growth is relatively flat so there is not much change in overall demand. Because this area has so much potential, most bio-lubricant suppliers have a range of products to meet the needs of this industry. The best estimate of the market share for bio-lubes in this sector today is about 1 %. Hydraulic oil bio-lubricants are used in some strategic areas such as national parks, military bases, national laboratories, golf course equipment, food services, some farm tractors and hydraulic elevators. It is the latter that is among the largest users of hydraulic fluids according to the section above.

9.9 Metal Working Fluid

Another successful application of a bio-lubricant came about in the rolling oil area of metal working for a bio-based rolling mill oil. The development is a success as it has found application four aluminum rolling mill operations around the world. They have also expanded the use of bio-based fluids in metal cutting and casting (mold release). It was established that the bio-based oil provides a cost savings compared with the traditional product. Part of this may be due to the carbon credits available in Europe. Vegetable oil-based cooling and cutting oils have also found success. The bio-lubricant showed substantial savings in coolant waste stream treatment and disposal, biocide use, tool savings and increased grinding throughput.

9.10 Chain Cutter Bar Oils

The bar and chain oil market is small, estimated at 2–3 million gallons per year. It is a total oil loss application as the lubricant serves to externally lubricate the bar and chain of the saw. All the expelled oil enters the environment, so it is another area where it makes sense to replace the mineral oil products with bio-lubricants. This is done in Europe but the present US demand for bio-lubes in this application is very limited. The major chain saw marketers indicate that the choice is up to the consumer as they fill the chambers after the saw is purchased. There are bio-lubes available, but the price is higher than the mineral oil counterparts. The choice made so far is usually for the lower cost product. It is estimated that only about 5 % of the bar and chain oil purchased is a biodegradable product. A positive indication in this area is that the major chain saw company has reportedly been receiving an increasing number of inquiries about bio-based lubricants for their products. This may well be due to the regulatory requirement to consider bio-based products for all lubricant applications.

9.11 Wire Rope Grease

The lubricants for wire rope applications were considered to be an area that had some regulations in effect. It was reported that Coast Guard regulations prohibited ships from conducting operations that produced a visible sheen in the waterways. This sheen developed because the wire rope, used for anchors, dredges and related applications, had grease lubricants on and in them. There were penalties imposed if these sheens were generated, except if biodegradable lubricants were employed. It is indicated that very little bio-based lubricant is used at present in this sector. It is up to the customer for the wire rope to specify that they want bio-lubricants in their rope, but this does not happen often as yet.

9.12 Railroad Lubricants

There are two types of rail-related lubricants that are total loss-type products, top-of-rail lubricants and gauge face greases. The top-of-rail lubricants are used to control friction; they are not there for wear control. Neither petroleum nor bio-based materials reportedly meet their needs. The product that they use is a water-based inorganic material that dries and controls friction. The market for this area is small, about 600,000 gallons per year. The gauge face grease application can use bio-based products.

9.13 Lubricant Condition Monitoring and Extended Drain

Conventional petroleum hydrocarbon based products are the predominant lubricants. In cases where operating conditions/parameters that exceed the performance characteristics of the petroleum based lubricants are taken into consideration, synthetic based lubricants are accepted as viable alternatives. It is anticipated that future advances and technological breakthroughs by using advanced condition monitoring or tribological systems would control lubricant degradation. The heart of the proactive maintenance program is the utilization of the machine and lubricant condition monitoring technologies. Proactive maintenance utilizes lubricant and machine service history and data accumulated through the utilization of the condition monitoring technologies to schedule repair, overhaul or replacement prior to machinery failure. The lubricant specification must qualify lubricants that meet specific performance characteristics to ensure long life in service. Lubricant condition monitoring and maintenance procedures/goals must be established which embrace "Cradle to Grave" ownership and service life of the lubricants and machinery assets. From the condition monitoring perspective, lubricant analysis continues to provide the earliest warnings of pending lubricant/machine anomalies. Any wear or distress in the machine will be evident using the wear particles monitoring technique (Tung et al. 2005).

In addition, many modern condition monitoring devices have been developed to set up in situ monitoring system to determine wear changes during the operation. The lubricant sample must be representative of the lubricant mass in order to better evaluate lubricant and machine condition. Continuing improvements in machinery design and lubricant extraction equipment/accessories are geared towards providing manufacturing with more efficient and consistent methods of obtaining lubricant samples. Typical lubricant monitoring technologies include: oil analysis, viscosity measurement, spectra-analysis, quantitative ferrography, analytical ferrography, Fourier Transform Infrared (FTIR). These monitoring techniques are very effective in detecting lubricant degradation and the onset of machine failures. The correlation between these fluid properties and the degradation mechanisms of fluids in a manufacturing environment can be made through these novel technologies.

Advanced and emerging developments in machinery and lubricant condition monitoring equipment continue to be an emerging. The capabilities of these monitoring methods in situ will be illustrated with real manufacturing data on lubricants of automotive. It is anticipated that future advances and technological breakthroughs using advanced condition monitoring in tribological systems will be beneficial for improving our manufacturing efficiency and controlling lubricant degradation.

9.14 Conclusion

Due to more stabilizing technologies, bio-lubricants have been increasingly successful in displacing mineral oil products in the areas of transformer fluids, elevator and other hydraulic fluids and metal rolling oils. Except for transformer fluids, the use of bio-lubricants only represents a small part of the total amount of lubricant used at this time. The present lack regulations in most parts of the world mandating the use of bio-lubricants in environmentally sensitive areas is major hurdle to be overcome of this area. The lack of regulatory pressures which mandate adoption and generally premium prices makes it difficult for suppliers to compete and for these areas to grow. However, this situation is changing as successes get more publicity. These vegetable-based lubricants are beginning to be considered as true alternatives to mineral oil products. The absence of bio-lubricant products from the major lubricant suppliers suggests that they have yet to decide that this area is large enough to justify their commercial participation at present. As a forecast, it is estimated that growth of bio-based lubricant products and fluids will be in the range of 5–8 % per year.

Revision Questions

1. Why is lubricant consumption important in face of emissions by vehicle engines?
2. Explain the technology used to stabilize bio based lubricants.
3. What drives the need for bio based lubricants on total loss applications?
4. Why is it that, these lubricants, have found major application in transformers, elevator hydraulics, metal rolling and chain cutting?
5. What are the limitations of use of bio based lubricants in crankcase application?

References

Anwar M, Kaushik RS, Srivastav M, Kumar M, Garg MO (2002) Basics: current base oil quality and technological options for its improvement—an Overview. Lubr Sci 14(4):425

ATC—Technical Committee of Petroleum Additive Manufacturers in Europe (2007) Lubricant additives and the environment. ATC Doc 49(Rev 1)

Pinchuk D, Pinchuk J, Akochi-Koblé E, Ismai AA, van de Voort FR (2010) Tribology and lubrication in extreme environments (two case studies). Thermal-Lube Inc.re, Québec

Tung SC, Paxton C, Liang F (2005) Overview and future trends of manufacturing lubrication and conditioning monitoring technologies. In: Proceedings of world tribology congress (WTC 2005), Washington (September 12–16, 2005)

Chapter 10
Recycling of Used Oil

Abstract Used oil recycling technology has undergone significant changes over the past decade. With the newly developed re-refining technologies it is possible to produce higher quality base oil compared with the traditional and old acid clay methods. Among other methods used are: solvent extraction (N-methyl-2-pyrrolidone, interline process, combined vacuum distillation and solvent extraction (Vaxon process), hydro–processing, combined thin film evaporation and hydro-finishing (CEP process), combined thermal de-asphalting and hydro finishing (Revivoil process). The majority of applied technologies are appropriate for re-refining of synthetic lubricating oils, which currently are replacing the conventional mineral lube oils due to their enhanced performance characteristics.

10.1 Introduction

Mineral oil components continue to form the most important foundation of lubricants and represent mixtures of different types of hydrocarbons with aliphatic hydrocarbons predominant and chemical additives. In contrast with mineral oil-based oils, that contain many different hydrocarbons, and nitrogen-, oxygen-, and sulfur-containing chemical derivatives of these hydrocarbons, synthetic base oils usually are prepared from a few well-defined chemical compounds, although in many cases based on petroleum minerals. Similar to mineral base oil, synthetic base oils generally cannot satisfy the requirements of high performance lubricants without modern additives. Moreover, some lube oil applications accept vegetable oils, which are biodegradable and contain long chain fatty acids in their composition.

When lubricating oils are used in service, they help to protect rubbing surfaces and promote easier motion of connected parts. In the process, they serve as a medium to remove high build-up of temperature on the moving surfaces. Further build up of temperature degrade the lubricating oils, thus leading to reduction in properties such as: viscosity, specific gravity, etc. Dirts and metal parts worn out

© Springer International Publishing Switzerland 2016
I. Madanhire and C. Mbohwa, *Mitigating Environmental Impact of Petroleum Lubricants*, DOI 10.1007/978-3-319-31358-0_10

from the surfaces are deposited into the lubricating oils. With increased time of usage, the lubricating oil loses its lubricating properties as a result of over-reduction of desired properties, and thus must be evacuated and a fresh one replaced. With the large amount of engine oils used, the disposal of lubricating oils has now become a major problem. In disposing used oil, many people use it as a dust cure; that is, for dust prevention. This method of disposal is in many ways unsatisfactory as the lead-bearing dust and run-off, constitute air and water pollution. Another method by which used oil is being disposed is by incineration. This method represents another poor use of such a valuable product, and the attendant emission of probably carcinogenous products, contribute to environmental pollution. Environmental considerations regarding the conservation of resources have maintained interest in the concept of recycling up to the present day. The reclamation of spent crankcase oils is now a subject of pressing national interest in some countries. On the other hand, pollution by used lubrication oils is recognized now to account for greater pollution than all oil spills at sea and off-shore put together (Udonne 2011).

Currently a number of automotive and industrial sources generate large amounts of used lubricating oils, which present a serious pollution problem. Due to high lube oil consumption various countries have designed their own systems for management of waste oils. As a result the re-refining industry has become an important industry in many countries. The ability to recycle waste oils is very closely linked to the oil's composition, level, type of contamination and of course economic aspects. Of these, only very few are primarily re-refiners, which recover lube oil for reuse. The others recycle waste oil by producing fuel for burning/energy recovery. Engine oil represents more than 70 % of the collectable waste oil and industrial oils comprise the balance of 30 %. About 35 % of the collected oil is re-refined into base oil; the remaining 65 % is burnt replacing coal (10 %), or used as heavy fuel oil (45 %) and unknown other products (10 %). It is also suspected that a substantial amount of used oil is lost or illegally burnt or dumped in the environment. Only should regeneration remain a priority due to its potential to conserve increasingly valuable natural resources but also due to the mitigation of other key environmental impacts, including reduction in emission of fine particles, carcinogenic risk potential and acidification (GEIR 2011). The basic principle of oil regeneration remains the same and utilizes many of the following basic steps (Udonne 2011):

- Removal of water and solid particles by settling.
- Sulphuric acid treatment to remove gums, greases, etc.
- Alkaline treatment to neutralize acid.
- Water washing to remove "soap".
- Clay contacting to bleach the oil and absorb impurities.
- Striping to drive off moisture and volatile oils.
- Filtering to remove clay and other solids.
- Blending to specification.

To avail adequate feedstock to recycling plants, waste oils have to be collected separately, where this is technically feasible; and waste oils are treated in accordance with waste hierarchy for protection of the environment and human health. Where this is technically feasible and economically viable, waste oils of different characteristics are not mixed with each other with other kinds of waste or substances, if such mixing impedes their treatment. Thus used oil regulations allow for four types of "recycling" which are reconditioning the oil on site to remove impurities; using the oil as a feedstock going into a petroleum refinery; re-refining the oil into a new base stock; or processing and burning the oil for energy recovery (Bremmer and Plonsker 2008).

10.2 Used Oil Disposal Challenges

Lubricating oils in service in automobiles and process industries become contaminated and lose their performance due to changes in some of their properties. The additives in the lubricants, say in engine oil, get depleted to such an extent that the lubricant cannot protect metal components any more. Therefore, such oils must be removed as used oil from the service as frequently as necessary. Issues on how to handle, and what to do with the used lubricants, are of serious concern to environmentalists, governments, industries and research scientists. Used lubricating oil disposal techniques of the past such as land filling, road oiling, track side foliage control, indiscriminate dumping, burning for energy, etc., create serious environmental problems. Many of these disposal techniques are severely restricted by current state and federal environmental regulations. The common disposal technique of used lubricating oil is burning for generation of energy. Burning and all other routes of disposals of used lubricating oils are un-economical and result in the wastage of resources.

A lot of research work is underway worldwide on degradation of lubricants, their analysis and recycling. The recycling of waste lubricating oils may be a suitable and economical alternative to burning and incineration. Different techniques have been developed for reclaiming and re-refining waste lubricants oils to either restore the original usefulness of the oil or clean the contaminated oils to a point that they can become suitable for sub-sequent use. Among these methods, reclaiming by heating and filtration, re-refining by introducing waste oil into crude oil refining streams, recycling by acid/clay treatment, thin film evaporation/clay contact finishing, **thin film evaporation/hydrotreats finishing**, etc., are now employed in different countries worldwide (Shakirullah et al. 2006).

Environmental concerns are associated with all those methods except the last which is usually the most expensive and needs a re-refining installation. Thin film evaporation methods are generally used in North America and Europe while the other methods, particularly the acid/clay technique are employed usually in the developing countries. This process has been employed for many years as the premier type of re-refining in which concentrated sulfuric acid is introduced to dehydrate waste lubricant oil. An acidic sludge is produced which is treated with clay. The acidic sludge and oily clay disposal is a matter of greater environmental concern. Changing away from the notorious acid/clay process to some form of distillation has been made in the past. Because of the environmental concerns of the acid methods, other methods based on solvents treatments were introduced which are economical because the solvents employed were recoverable. These methods will be helpful to control effectively environmental pollution due to acid reclamation processes.

Recovered lubricating oils are characterized to establish physico-chemical properties in order to investigate the similarities and differences with virgin oil. This comparison of properties will be helpful in identifying an effective, reliable and more appropriate environmentally friendly method among the methods developed. By addition of some additives, certain properties of the recovered lubricants may be further improved to render them as excellent as virgin lubricants.

10.3 Waste Oil Recycling Basics

Due to the increasing necessity for environmental protection and more and more strict environmental legislation, the disposal and recycling of waste oils has become very important. The recycling of waste lubricating oils can be accomplished with the following different methods (Kupareva et al. 2013):

- used oil reprocessing,
- reclamation and
- re-refining.

The products of used oil reprocessing and reclamation are lubricating fluids with low quality requirements and heating and fuel oils, respectively, while re-refining of the used oil leads to production of valuable base oils with quality comparable with virgin base oils. The principle of rerefining waste oils utilizes the following four steps:

- dewatering and defueling,
- de-asphalting,
- fractionation and
- finishing process.

The re-refining process is depicted schematically in Fig. 10.1. The majority of collected lube oils is automotive oil. The main functions of lubricating oils include reducing friction, carrying away heat, protecting against rust, protecting against wear and removing contaminants from the engine. All the above-mentioned properties are obtained owing to the package of additives carefully incorporated in appropriate quantities.

The separation of the lubricant base oils from additives, asphalts and other contaminants contained in the used oil has been performed traditionally by distillation

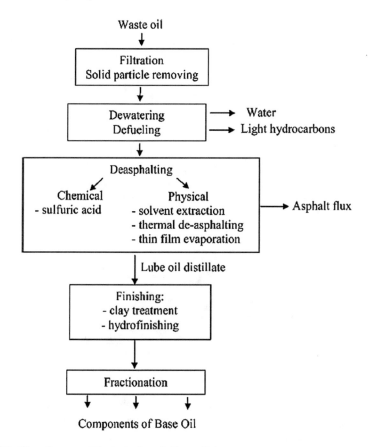

Fig. 10.1 Flow diagram of the typical used oil re-refining process

and acid/clay treatment, however, these traditionally applied technologies have been discarded due to the following problems:

- Equipment fouling, in particular in the distillation equipment, reducing the operational time of the plant.
- Thermal cracking reactions reducing the yield and leading to low quality base oils (color, odor, instability etc.), which is difficult to improve in the final treatment.
- Environmental difficulties due to the acid-clay waste disposal, emission of unpleasant odors and contaminated water.

To avoid thermal cracking and fouling, several processes have been developed, employing high vacuum (about 1 mbar) in the distillation, thus separating asphalts and additives at lower temperatures (falling film evaporator, short path distillation, etc.). As an alternative to the separation of asphalts and additives by vacuum distillation, extraction processes, using liquid solvents, have been developed (solvent

Table 10.1 Properties of used oils, intermediate products during re-refining

Properties	Used oil	Feed of vacuum distillation	Feed of hydrogenation		Rerefined oil	
			Light	Heavy	Light	Heavy
Water content (%)	10	<1	<1	<1	–	–
Density at 15 °C (kg m^{-3})	~900					
Kinematic viscosity at 40 °C (mm^2 s^{-1})	80					
Flash point (°C)	>60	215	154	193	182	210
Total acid number (mg KOH g^{-1})	<2.4	4.0	1.5	0.5	0.01	0.01
Sulfur content (mg kg^{-1})	4000	4000	3000	3000	600	600
Zinc content (mg kg^{-1})	650	650	<1	<1	<1	<1

Table 10.2 API base oil categories

Group	Sulfur (wt%)	Saturates (wt%)	Viscosity index
I	>0.03		
II	≤0.03		
III	≥0.03		
IV	All polyalphaolefins (PAOs)		
V	All others not included in Groups I, II, III or IV		

de-asphalting). These processes operate at near ambient temperatures, thus avoiding to a large extent the equipment fouling problems and the cracking of asphalts, additives and breakdown products since these are separated before distillation of the lubricant bases. Table 10.1 demonstrates the key properties of used oils as well as the property changes that occur after re-refining.

The properties of re-refined used lube oils are similar to the fresh ones. The American Petroleum Institute (API) categorized base oils by composition, as shown in Table 10.2. Modern regeneration technologies allow to produce premium quality base oils belonging to at least Group I according to the API base oils classification. Under more severe or solvent finishing conditions, Group II base oils could be obtained.

10.4 Acid-Clay Re-refining Process

The acid-clay re-refining process was the first regeneration process commercialized by many companies, where large amounts of sulphuric acid and clay were used to treat the waste oils. Acid-clay re-refining processes have been widely used by re-refining facilities. Used oil is pre-treated (preflash or vacuum distillation) for separation of water and light hydrocarbons. Concentrated sulfuric acid (10–15 wt%)

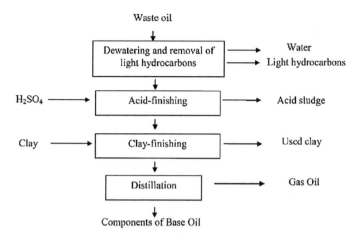

Fig. 10.2 Block flow diagram of the acid-clay re-refining process

is added into dehydrated waste oil, wherein the foreign substances (e.g. additives and sulfides) will form sludge, enabling deposition within 16–48 h which is thereafter separated from the waste oil. The impurities such as colloids, organic acids and waxy substances are removed by clays (porcelain clay or aluminum silicate). Filtered oil is distilled to produce base oils with various characteristics and gas-oil. Figure 10.2 provides the flow chart of the process.

The base oil obtained has low quality with a lubricating yield of 62–63 %on dry basis. The product oils are dark in color and tend to have a noticeable odor. Moreover, the products have from 4 to 17 times higher content of polycyclic aromatic hydrocarbon (PAH) than virgin oils. While the technology has relatively low capital costs and simplicity of operations, as well as allowing production of acceptable, although sub-standard, base oils, it also generates acid tar, oil saturated clay, and other hazardous waste by-products. Under increasing environmental pressure this technology has been banned in most countries including many developing countries.

10.5 Hylube Process

The Hylube process allows production of mainly base oils. The Hylube process is a proprietary process developed by Universal Oil Products (UOP) for the catalytic processing of used lube oils into re-refined lube base stocks for re-blending into saleable lube base oils. This is the first re-refining process in which as received used oil is processed, without any pretreatment, in a pressurized hydrogen environment. A typical Hylube process feedstock consists of a blend of used lube oils containing high concentrations of particulate matter such as iron and spent additive contaminants such as zinc, phosphorous and calcium.

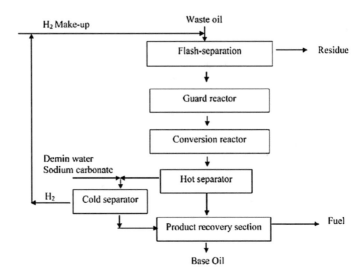

Fig. 10.3 Block flow diagram of the Hylube process

A simplified block diagram of the process is shown in Fig. 10.3. The first part of the process involves separation of the lube range and lighter components of the feed from the non-distillable residue portion. After the separation step the light feed is flowed through the so-called 'guard' reactor where metal-containing compounds and other impurities are accumulated in the large pore size catalyst.

The treated feed is hydrogenated in the main reactor before the second separation step. The Hylube unit operates with reactor section pressures of 60–80 bar and reactor temperatures in the range 300–350 °C. As the feed is processed in hydro finishing reactors, contaminants are removed and the quality of the lube base oil is rejuvenated and enhanced. In addition to converting hetero-atoms such as sulfur and nitrogen, the catalyst is able to increase the viscosity index via saturation of multi-ring aromatic compounds. After the hydrogenation, products are stripped and separated in the fractionation tower to gasoline, petroleum, gas oil and base oil fractions. Light ends from the high temperature separator are blended with sodium carbonate and flowed to the low temperature separator, where the waste water is settled and separated. The hydrogen rich vapor from the cold separator is scrubbed, compressed, reheated and returned to the mixer.

The hydrocarbon liquids collected in the separators are sent to the product fractionation section where the products are separated into various cuts to meet the desired lube oil viscosity grades. The processed feedstock is converted into a wide boiling range hydrocarbon product, which is subsequently fractionated into neutral oil products of different viscosity to be used for lube oil blending. Due to hydrogenation the properties of three different base oil products are the same as the properties of fresh Group II base oils Table 10.3.

The Hylube process achieves more than 85 % lube oil recovery from the lube boiling range hydrocarbon in the feedstock.

Table 10.3 Properties of base oil products of Hylube-process

Properties	Base oils		
	Light grade (PUR-75)	Medium grade (PUR-160)	Heavy grade (PUR-300)
Density at 15 °C (kg m^{-3})	850	855	860
Flash point (°C)	190	215	228
Pour point (°C)	−12	−12	−12
Kinematic viscosity at 40 °C (mm^2 s^{-1})	13.5	29.5	58
Kinematic viscosity at 100 °C (mm^2 s^{-1})	3.19	5.2	8.4
Viscosity index	100	115	116
Sulfur content (ppm)	100	100	100
Saturates (%wt)	≫90	≫90	≫90

10.6 Mineralöl Raffinerie Dollbergen (MRD) Solvent Extraction Process Using N-Methyl-2-Pyrrolidone

This technology has been processing and recycling used oil and oily liquids since 1955. The 'Enhanced Selective Refining' process uses solvent N-methyl-2-pyrrolidone (NMP), which is commonly used in the petroleum refining industry. NMP is a powerful, aprotic solvent with low volatility, which shows selective affinity for unsaturated hydrocarbons, aromatics, and sulfur compounds. Due to its relative non-reactivity and high selectivity, NMP finds wide applicability as an aromatic extraction solvent in lube oil re-refining. The advantages of NMP over other solvents are the non-toxic nature and high solvent power, absence of azeotropes formation with hydrocarbons, the ease of recovery from solutes and its high selectivity for aromatic hydrocarbons. Being a selective solvent for aromatic hydrocarbons and PAH, NMP can be used for the re-refining of waste oils with lower sludge, carbonaceous particles and polymer contents, such as waste insulating, hydraulic and other similar industrial oils.

The MRD solvent extraction process uses the liquid–liquid extraction principle. Figure 10.4 provides the flow chart of the process. Vacuum distillates from the flash distillation are used as feed. These distillates are processed in a production cycle which can be adjusted to the quantity to be processed. Before the distillate enters the extraction column, any residues of dissolved oxygen in the distillate are removed in an absorber using steam.

Thereafter the distillate is sent to the bottom part of the extraction column. As the distillate rises, undesirable aromatic hydrocarbons and other contaminants are separated out by the counter-flowing heavier solvent, N-methyl pyrrolidone, which is fed in at the top of the extraction column. The solvent containing raffinate phase leaves the extraction column at the top and is routed to the downstream raffinate recovery section consisting of a distillation and a stripping column where the solvent is removed. The extract phase is continuously withdrawn from the bottom of the extraction column, cooled down to a defined

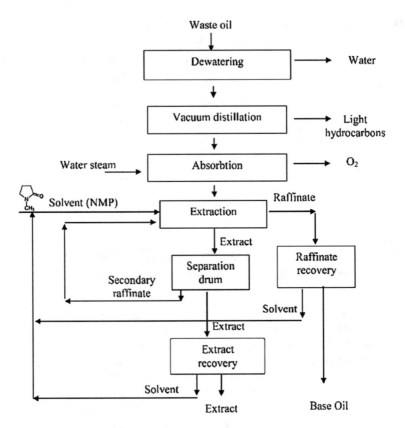

Fig. 10.4 Block flow diagram of the MRD Solvent extraction process

temperature and separated in a separation drum from the separated secondary raffinate. The latter is returned to the extraction column in order to optimize the process yield. The extract phase from the secondary separation drum is sent to the extract recovery section where the solvent is removed. The extract recovery section also consists of a distillation and a stripping column. The resulting extract is routed to the off plot intermediate storage tank and used within the refinery as an energy carrier or mixing component for heavy oil. The dry solvent separated in the distillation columns of the raffinate and extract recovery sections is returned to the solvent tank. The moist solvent separated in the stripping columns of the raffinate and extract recovery sections is returned to the solvent drying column, where excess water is removed.

The average base oil yield within the process is about 91 %. The base oils produced have high quality (Table 10.4). The process is characterized by optimized operating conditions which allow elimination of toxic polyaromatic compounds from the re-refined base oil and preservation of the synthetic base oils like polyalphaolefin (PAO) or hydrocracked oils, which are increasingly present in used oils.

Table 10.4 Properties of base oil products of MRD solvent extraction process

Properties	Base oils		
	Light grade (KS 100)	Medium grade (KS 150)	Heavy grade (KS 200)
Density at 15 °C (kg m^{-3})	852–856	857–860	860–865
Flash point (°C)	>220	>230	>230
Pour point (°C)	−12	−9	−9
Kinematic viscosity at 40 °C (mm^2 s^{-1})	22–26	32–36	40–46
Kinematic viscosity at 100 °C (mm^2 s^{-1})	4.4–4.9	5.5–5.6	6.4–7.1
Viscosity index	108–112	110–115	110–115
Sulfur content (wt%)	≤0.25	≤0.25	≤0.25

10.7 Vaxon Process

This process contains chemical treatment, vacuum distillation and solvent refining units. The advantage of the Vaxon process is the special vacuum distillation, where the cracking of oil is strongly decreased (Fig. 10.5).

Used oil is chemically treated with alkali-hydroxides (sodium and potassium hydroxide) for removal of chlorides, metals, additives and acidic compounds. The impurities can be bonded with asphaltene molecules by these reactants, therefore these impurities can be easily separated from the oil. After the chemical treatment the oil is separated to light products, catalyst, base oil and residue. The feed is distilled to two parts by a cyclonic column. Because of the formation of tangentially flowed thin film the light hydrocarbons are easily and quickly distilled. The evaporated lighter part, consisting of light hydrocarbons (gas, diesel fuel) and water, is condensed in the upper part of the chamber, from where it is separated. The heavier oil part, circulating in the bottom, is heated, thus decreasing heat transfer and decreasing coke formation in the chamber.

The process can be carried out in several evaporators at various temperatures and pressure (i.e. from 160 to 345 °C and vacuum from 400 to 5 mbar) allowing separation of several oil cuts. Oil from the distillation can contain some undesirable components and should be additionally treated. The polycyclic aromatic hydrocarbons are separated by solvent refining with polar solvents (dimethyl-formamide, N-methyl-2-pirrolidone, etc.). This is carried out in a multi-stage extractor, which is followed by solvent recovery from both phases. The treated raffinate can be additionally distilled to obtain various base oil cuts with yields of 65–70 %. The polycyclic aromatic hydrocarbons, which concentrate in the extract, are used for heat energy production or as bitumen blending component. The chemical final stage does not, however, allow the production of high quality base oils; although in Spain the Catalonia refinery produces base stocks accepted by an original equipment manufacturer (OEM).

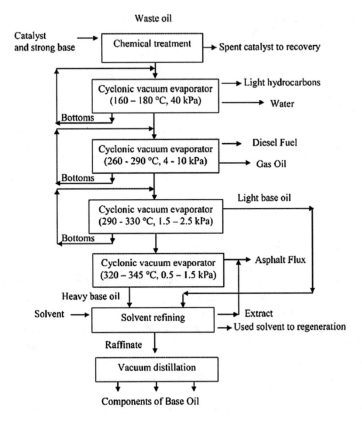

Fig. 10.5 Vaxon process block flow diagram

10.8 CEP Process

The process combines thin film evaporation and hydro processing (Fig. 10.6).

The used oil is chemically pretreated to avoid precipitation of contaminants which can cause corrosion and fouling of the equipment. The pre treating step is carried out at temperatures from 80–170 °C. The chemical treatment compound comprises sodium hydroxide, which is added in a sufficient amount to give a pH about 6.5 or higher. The pre-treated used oil is first distilled for separation of water and light hydrocarbons. Water is treated and sent to a waste water treatment facility. Light hydrocarbons are used at the plant as fuel or sold as a product. Thereafter, free-of-water oil is distilled under high vacuum in a thin film evaporator for separation of diesel fuel, which can be used at the plant or sold as fuel. Heavy materials such as residues, metals, additive degradation products, etc. are passed to a heavy asphalt flux stream. The distillate is hydro purified at high temperature (315 °C) and pressure (90 bar) in a catalytic fixed bed reactor. This process removes nitrogen, sulfur, chlorine and oxygenated organic components.

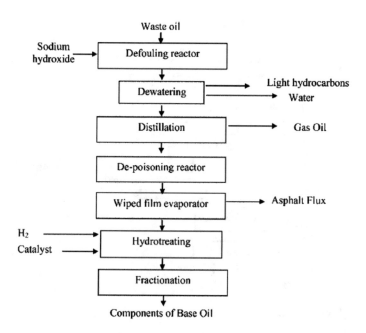

Fig. 10.6   Block flow diagram of the CEP process

In the final stage of the process, three hydrotreating (Hydrofinishing) reactors are used in series to reduce sulfur to less than 300 ppm and to increase the amount of saturated compounds to over 95 %, in order to meet the key specifications for API Group II base oil (Table 10.5). The final step is vacuum distillation to separate the hydrotreated base oil into multiple viscosity cuts in the fractionator. Hydro processing technology is one of the most widely used distillation processes to eliminate undesirable components such as sulfur, nitrogen, metals or unsaturated hydrocarbons. The yield of base oils is about 70 %.

Table 10.5   Properties of base oil products of CEP process

Properties	Medium grade (base oil—N150)
Density at 15 °C (kg m^{-3})	840–860
Flash point (°C)	>200
Pour point (°C)	<–9
Kinematic viscosity at 40 °C (mm^2 s^{-1})	26–32
Kinematic viscosity at 100 °C (mm^2 s^{-1})	5–6
Viscosity index	>110
Sulfur content (ppm)	<300
Saturates (%wt)	>95

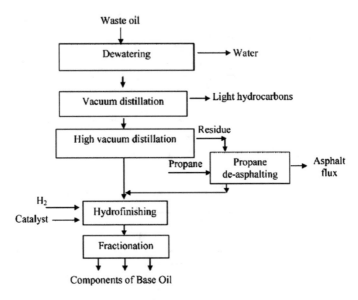

Fig. 10.7 Flow diagram of the Cyclon process

10.9 Cyclon Process

The process flow diagram of the Cyclon process is illustrated in Fig. 10.7. Used oils taken from storage tanks are dewatered and the light hydrocarbons are removed by distillation. The heavier fraction is sent to high vacuum distillation, where the majority of base oil components are evaporated from the heavy residue. The oils in the residues are extracted with propane in the de-asphalting unit and sent to the hydro processing unit where the other oils are processed. Then they are treated with hydrogen and fractionated based on the desired base oil features.

10.10 Snamprogetti Process/IFP Technology

The Snamprogetti process combines vacuum distillation and hydrogenation including a propane extraction step before and after vacuum distillation (Fig. 10.8). The extraction technology is similar to the one carried out in crude oil refineries to separate out asphaltenes. In the first stage of the Snamprogetti technology, the light hydrocarbons and water are removed by atmospheric distillation.

In the second stage, all the impurities picked up by the engine oil, including the additives and partly degraded polymers, are removed by extraction with propane in the temperature range 75–95 °C in a propane de-asphalting section (PDAI).

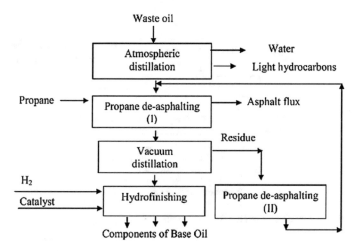

Fig. 10.8 Snamprogetti process block flow diagram

After stripping the propane, the oil is heated again and vacuum distilled at a temperature of 300 °C. In this stage lubricating bases having lower viscosity and free of impurities are separated. The vacuum residue is then heated to 300–450 °C under adiabatic conditions and sent to the second extraction stage (PDA II) in which metal content and asphaltic components are further reduced. After extraction, propane is stripped and recycled in the process. The base oil cuts from the vacuum residue (bright stock) are finally hydrogenated to improve the color and to increase the oxidation stability of the base oils.

10.11 Revivoil Process

Revivoil process is made up of three key sections: pre-flash, thermal de-asphalting and hydro finishing (Fig. 10.9). The filtered used oil from storage tanks is heated to 140 °C and then distilled in a pre-flash column where the water and light hydrocarbons are separated. The de-hydrated oil is distilled at 360 °C in vacuum in a thermal de-asphalting unit (TDA), where the oil is separated from substances that can enhance fouling in an intermediate tank. The asphaltic and bituminous products remain at the bottom and three side cuts of different viscosities are obtained at the same time. Intermediate gas oil is collected from the top of the column.

For improvement of the product quality, oil cuts after TDA are treated with hydrogen over the catalyst. The hydro finishing process starts in a fired heater where the oil and hydrogen are heated to 300 °C. They are then sent to a reactor containing a catalyst favoring hydrogenation of the unsaturated compounds, as well as sulphur and nitrogen containing compounds. The reactor effluent is then separated into two phases, the vapor phase and the liquid phase; the first one is

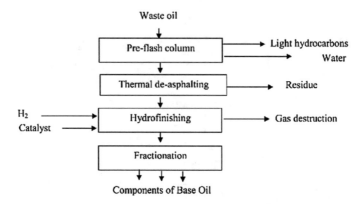

Fig. 10.9 Block flow diagram of the Revivoil process

Table 10.6 Properties of base oil products of Revivoil process

Properties	Base oils		
	Light grade	Medium grade	Heavy grade
Density at 15 °C (kg m^{-3})	852	853	858
Kinematic viscosity at 40 °C (mm^2 s^{-1})	16.5	30.6	55.2
Kinematic viscosity at 100 °C (mm^2 s^{-1})	3.6	5.3	7.8
Viscosity index	101	106	107
Sulfur content (wt ppm)	<300	<300	<300

washed with water to remove the chlorine and sulphur compounds, the second one is stripped with steam to eliminate the most volatile compounds and restore the flash point. The water contained in the oil after stripping is then removed in a vacuum dryer. The yield of base oils from the Revivoil process is about 72 %. According to the operating parameters of hydro finishing, the final base oil quality can be upgraded until the amounts of sulphur and saturated compounds fulfil the API Group II requirements (Table 10.6).

10.12 Latest Used Oil Re-refining Technologies

As people try to mitigate the damaging effect of used oil on the environment, there are now several new technologies, such as thin film evaporation (TFE), including combined TFE and clay finishing, TFE and solvent finishing, TFE and hydro finishing, thermal de-asphalting (TDA), TDA and clay finishing, and TDA and hydro finishing. In addition, solvent extraction and hydro finishing are being developed by means of hydro finishing after the solvent de-asphalting process. Various technologies vary in terms of operating and capital costs, quality of feedstock and products obtained. The technologies described can be divided into the following groups:

- Solvent extraction process
- Hydro processing

Combined processes:

- Vacuum distillation or thin film evaporation and finishing process (solvent extraction or chemical treatment)
- Thin film evaporation and hydro finishing
- Thermal de-asphalting and hydro finishing
- Solvent extraction and hydro finishing

Thin film evaporation technology includes a rotating mechanism inside the evaporator vessel which creates high turbulence and thereby reduces the residence time of feed-stock oil in the evaporator. This is done in order to reduce coking, which is caused by cracking of the hydro carbons due to impurities in the used oil. Cracking starts to occur when the temperature of the feedstock oil rises above 300 °C. However, any coking which does occur will foul the rotating mechanism and other mechanisms such as tube-type heat exchangers are often found in thin film evaporators.

Solvent extraction processes are widely applied to remove asphaltic and resinous components. Low molecular weight hydrocarbons as solvents selectively dissolve the undesired aromatic components, the extract, leaving the desirable saturated components, especially alkanes, as a separate phase, the raffinate. Liquid propane is by far the most frequently used solvent for de-asphalting residues to make lubricant bright stock, whereas liquid butane or pentane produces lower grade de-asphalted oils more suitable for feeding to fuel-upgrading units. The liquid propane is kept close to its critical point and, under these conditions, raising the temperature increases selectivity. A temperature gradient is set up in the extraction tower to facilitate separation. Solvent-to-oil ratios are kept high because this enhances rejection of asphalt from the propane/oil phase. Countercurrent extraction takes place in a tall extraction tower, of the type shown in Fig. 10.10. The solvent chosen should meet the following requirements: maximum solubility for the oils and minimum solubility for additives and carbonaceous matter; ability to be recovered by distillation. New plant units increasingly use N-methyl pyrrolidone because it has the lowest toxicity and can be used at lower solvent/oil ratios, saving energy.

For several years, catalytic hydro treatment stood out as the modem and successful refining treatment from the point of view of the yield and quality of the finished products. Hydro processing is more often applied as a final step in the re-refining process in order to correct problems such as poor color, oxidation or thermal stability, de-mulsification and electrical insulating properties. A simplified flow diagram of a hydro-finishing plant is shown in Fig. 10.11. Oil and hydrogen

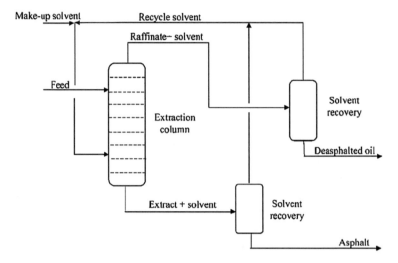

Fig. 10.10   A block flow diagram for solvent extraction

Fig. 10.11   Flow diagram of hydro-finishing

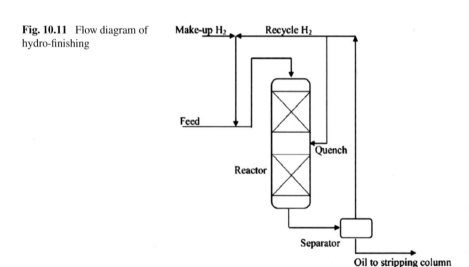

are pre-heated and the oil allowed to trickle downwards through a reactor filled with catalyst particles where hydrogenation reactions take place. The oil product is separated from the gaseous phase and then stripped to remove traces of dissolved gases or water.

The following reactions can be operative: hydro refining reactions with the objective of removing hetero elements and to hydrogenate olefinic and aromatic compounds, and hydro conversion reactions aiming at modifying the structure of hydrocarbons by cracking and isomerization. 35 Hydro treatment catalysts are made of an active phase constituted by molybdenum or tungsten sulfides as well as by cobalt or nickel on oxide carriers.

The technologies applying hydro processes require relatively high investments compared with others. However, depending on the technology adopted, the total cost might be lower than in solvent extraction process due to the high operating costs to make up for the solvent losses. On the other hand, solvent extraction and chemical treatment processes do not require catalyst regeneration. Moreover, it is not necessary to establish a hydrogen gas supply facility in these methods which in addition reduces a risk concerning operation safety.

The technologies applying hydro processing obtain product oils with the highest quality independently of the quality of feedstock. Thus technologies such as Hylube, CEP, Revivoil, Snamprogetti and Cyclon produce high-quality base oils, which fulfill the API Group II and even II$^+$ requirements. In terms of the nature of the feedstock, some synthetic oils which have enhanced performance characteristics and currently are replacing conventional mineral lube oils, can be regenerated along with mineral oils.

The drawback of solvent extraction is dependence of the product quality on hydrocarbon feedstock composition. The high quality base oils API Group II/II$^+$ can be obtained by solvent extraction methods only when the reaction mixture is homogeneous. The majority of current solvent extraction technologies (Interline, Snamprogetti, Cyclon) use propane as a solvent, which has a lower selectivity to undesired species compared with MRD. Moreover propane requires high solvent–oil ratios which increase energy consumption. From economical and technological points of view replacing propane with MRD looks feasible.

10.13 Recycling of Waste Engine Oils Using Acetic Acids

In 2004 the global lubrication market was estimated to be roughly 37.4 million tons of lubricants as given in Fig. 10.12 by application. This figure illustrates how automotive and industrial lubricants are the most prevalent. Automotive lubricants is the most commonly used liquid lubricants with about 53 % of the volume while *engine oils* constituted the bulk of this amount ahead of automatic transmission fluids, gearbox fluids, brake fluids, and hydraulic fluids. Industrial lubricants amount to 32 % and were composed of hydraulic oils, other industrial oils, metalworking fluids, greases, and industrial gear oils (Reeves and Menezes 2016).

From above it follows that most waste lubricating oil handled is engine oil which would have gone through degradation and contaminated by metals, ash, carbon residue, water, varnish, gums, and other contaminating materials, in addition to asphaltic compounds which result from the bearing surface of the engines. Thus these oils must be changed and removed from the automobile after a few thousand kilometers of driving because of stress from serious deterioration in service.

If discharged into the land, water or even burnt as a low grade fuel, this may cause serious pollution problems because they release harmful metals and other

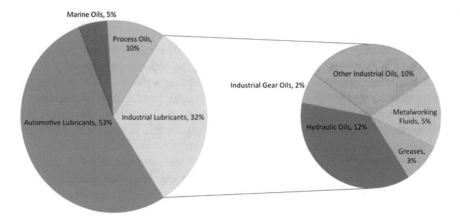

Fig. 10.12   Worldwide lubricating oils consumption

pollutants into the environment. Waste engine oil may cause damage to the environment when dumped into the ground or into water streams including sewers. This may result in groundwater and soil contamination.

A recommended solution for this issue is the recovery of the lubricating oil from the waste oil. Recycling processes using nontoxic and cost effective materials can be an optimum solution. The conventional methods of recycling of waste engine oil either requires a high cost technology such as vacuum distillation or the use of toxic materials such as sulfuric acid. These methods also produce contaminating by-products which have highly sulfur levels.

Acid-clay has been used as a recycling method for used engine oil for a long time. This method has many disadvantages, as it also produces large quantity of pollutants, and is unable to treat modern multi grade oils and it is difficult to remove asphaltic impurities. Solvent extraction has replaced acid treatment as the method of choice for improving the oxidative stability and viscosity/temperature characteristics of base oils. The solvent selectively dissolves the undesired aromatic components (the extract), leaving the desirable saturated components, especially alkanes, as a separate phase (the raffinate).

Membrane technology is another method for regeneration of used lubricating oils. The process is carried out at 40 °C and 0.1 MPa pressure. The process is a continuous operation as it removes metal particles and dusts from used engine oil and improves the recovered oils liquidity and flash point. Despite the above mentioned advantages, the expensive membranes may get damaged and fouled by large particles.

Thus, process recycling of waste engine oils by treating them using acetic acid, was developed and found to have comparable results with some of the conventional methods discussed above. The recycling process takes place at room temperature and atmospheric pressure. The process for recycling is simple, as it only requires mixing, settling, centrifugation and finally mixing with kaolinate.

This gives the recycled oil the potential to be reused in cars' engines after adding the required additives. The advantage of using the acetic acid is that it does not react with base oils. It has been shown that base oils and oils' additives are slightly affected by the acetic acid. Upon adding 0.8 volume percentage (%) of acetic acid to the used oil, two layers were separated, a transparent dark red colored oil and a black dark sludge at the bottom of the container. The use of glacial acetic acid for recycling used engine oils provides a lower cost process in comparison with the conventional methods due to the low cost of the acid and the moderate conditions of the process. Lower amount of additives may be required for the base oil recycled by acetic acid-clay method due its low reactivity with the used oil. This new process of recycling of used engine oil did not emit poisonous gases like sulfur dioxide to the atmosphere. The recycled oil obtained by this method has been shown to have potential for reuse as an engine lubricant (Hamawand et al. 2013).

10.14 Conclusion

From an environmental perspective the regeneration of waste oils makes an extremely positive contribution. Not only does it alleviate the significant environmental burden of the primary production of lubricants and represent the largest and most advantageous recovery option, (ensuring the proper collection of waste oils as opposed to incineration which attracts the unwanted mixing of wastes) but it also has other environmental benefits, for example modern re-refined products fulfill the needs of motor vehicle OEMs, which need high quality products with low sulfur, aromatic and phosphorus content in order to respond to the Kyoto Protocol. Responding to market demand the European regeneration industry has in recent years made significant technological advances in terms of the quality of regenerated products, production efficiency and environmental impact, putting them on an equal footing with virgin base oils and providing an important economic alternative (GEIR 2011).

Under increasing environmental pressure acid clay treatment, which was the first oil regeneration process used, was substituted in the majority of European countries with new technologies based on solvent extraction and hydro processing. Leading industrial processes employ different technologies, such as combined thermal de-asphalting and hydro finishing (Revivoil), solvent extraction (MRD process, Interline), solvent extraction and hydrofinishing (Cyclon, Snamprogetti), thin film evaporation and different finishing process (Ecohuile, Vaxon, CEP) and hydroprocessing (Hylube). Thus, currently the most attractive method for re-refining used oil could be a combination of MRD-solvent extraction and hydro finishing, since application of the hydro finishing step gives product oils of high quality independent of the feedstock nature. The MRD-solvent extraction process allows reduced catalyst poisoning without any alkaline treatment of the used oil. Absence of a need to apply alkali agents makes it possible to regenerate synthetic and semi-synthetic oils along with mineral oils. The composition of hydro processing catalysts could be optimized to increase catalyst stability and achieve the highest oil conversion and yield.

Revision Questions

1. *What environmental challenges are posed by used oil?*
2. *Which two major basic processes are involved in used oil cycling?*
3. *Outline the recycling process to point of base oil, indicate why mostly it is considered inferior?*
4. *Name any five common re-refining processes available?*
5. *What advances have been made to enable synthetic and semi-synthetic to be processes together with mineral oils?*

References

Bremmer BJ, Plonsker L (2008) Bio-based lubricants: a market opportunity study update. United Soyabean Board, Omni Tech International, Ltd

GEIR—Groupement Européen de l'Industrie de la Régénération (2011) An environmental review of waste oils regeneration: why the regeneration of waste oils must remain an EU policy priority. The Re-refining Industry Section of UEIL, Brussels

Hamawand I, Yusaf T, Rafat S (2013) Recycling of waste engine oils using a new washing agent. Energies 6:1023–1049. doi:10.3390/en6021023, http://www.mdpi.com/journal/energies

Kupareva A, Mäki-Arvela P, Yu Murzin D (2013) Technology for rerefining used lube oils applied in Europe: a review. J Chem Technol Biotechnol 88:1780–1793. doi:10.1002/jctb.4137, http://www.wileyonlinelibrary.com

Reeves CJ, Menezes PL (2016) Advancements in eco-friendly lubricants for tribological applications: past, present, and future: ecotribology research developments—materials forming, machining and tribology, vol 2. Springer, Switzerland, pp 41–61. doi:10.1007/978-3-319-24007-7_2, http://www.springer.com/978-3-319-24005-3

Shakirullah M, Ahmad I, Saeed M, Khan MA, Rehman H, Ishaq M, Shah AA (2006) Environmentally friendly recovery and characterization of oil from used engine lubricants. J Chin Chem Soc 53:335–342

Udonne JD (2011) A comparative study of recycling of used lubrication Oils using distillation, acid and activated charcoal with clay methods. J Pet Gas Eng 2(2):12–19. http://www.academicjournals.org/JPGE (Academic Journals)

Chapter 11
Environment and the Economics of Long Drain Interval

Abstract The majority of lubricants used in the world today result in environmental pollution through total-loss applications, spillage, evaporation and mishandling. As a way to reduce this environmental damage, new lubricants that are rapidly biodegradable and ecologically non-toxic are being developed. This chapter dwells on the need to highlight the use of environmentally friendly lubricants and their constituents, with particular emphasis on their environmental benefits, applications, the limits to their use, their technical performance characteristics and related cost aspects. Vegetable and ester based lubricants are compared with conventional mineral-oil based materials.

11.1 Introduction

Petroleum based lubricants had a significant cost advantage over bio-lubricants and so petroleum has been the base oil of economic choice. The use of an 'environmentally acceptable' lubricant (EAL) is often a requirement when operating in environmentally sensitive areas such as inland waterways, flood defenses, harbors, water catchment areas, forests and ski slopes. Whilst the best way of protecting the environment is through good housekeeping and maintenance, accidents can happen and even an apparently small leak of, say, 1 drop/s can lead to the release of up to 200 L of lubricant in a month (Battersby 2005).

Recently, bio-based lubricants have begun to seek prominence for their environmental friendliness and superior tribological properties. The current trend in the lubrication industry is to develop more bio-based lubricants due to estimates indicating that nearly 50 % of all lubricants sold worldwide pollute the environment, through spillage, evaporation, and total loss applications. An example of lubrication pollution is that of the diesel engine particulate emissions, where approximately one-third of the engine oil vaporizes thus polluting the atmosphere. The large quantity of lubricant loss into the environment is the reason behind the development of eco-friendly lubricants (Reeves and Menezes 2016).

© Springer International Publishing Switzerland 2016
I. Madanhire and C. Mbohwa, *Mitigating Environmental Impact of Petroleum Lubricants*, DOI 10.1007/978-3-319-31358-0_11

It is estimated that in excess of half of lubricants sold worldwide end up polluting the environment through total loss applications, spillages, evaporation and mishandling. This environmental awareness has led to the development of environmentally harmless lubricants in the early 1980s. The additive industry has supported these developments with environmentally harmless lubricant additive packages. Where the environmentally harmless means that the lubricants are rapidly biodegradable and ecologically non-toxic. The major oil losses to the environment are given in Table 11.1.

There are many national and international schemes for defining an EAL but most stipulate that the lubricant must have:

- high biodegradability (rapid removal from the environment i.e. fate)
- low toxicity to plants and animals (minimal impact on the environment i.e. effects)
- minimum technical performance (work as a lubricant i.e. function)

Currently, EAL are predominantly found in total loss applications (e.g. two-stroke oils, chain saw oils, concrete mould release oils, wire rope lubricants, greases for railway points and wheel flanges); or where there is high risk of leaks (e.g. hydraulic fluids in mobile equipment). An example of the requirements for this latter group of EAL is shown in Table 11.2.

Although automotive products constitute approximately half the global annual consumption of lubricants, the use of EAL has been limited to concept vehicles demonstrating 'green technologies'. The competing demands of ever stricter

Table 11.1 Lubricants in the environment (for Germany)

Loss to environment	Percentage (%)
Collected as waste oil	47
Re-used or disposed of by customers	11
Burn in engines	6
Lost in circulation systems	28
Total loss lubrication	8

Table 11.2 ISO 15380:2002 requirements for environmentally acceptable hydraulic fluids based on synthetic esters

	Characteristic	Requirement
Fate	Biodegradation to CO_2	≥ 60 % within 28 days
Effects	Concentration which kills 50 % test fish after 96 h	≥ 100 mg/l
	Concentration which inhibits growth of algae by 50 % after 48 h exposure	≥ 100 mg/l
	Concentration which inhibits respiration of bacteria by 50 % after 48 h exposure	≥ 100 mg/l
Function	Kinematic viscosity, corrosion protection, foaming, demulsibility, elastomer compatibility, oxidative stability, load carrying and antiwear properties	Meet minimum VDMA performance requirements

Table 11.3 Typical base fluids for EAL against mineral oil base

Performance	Canola oil	Synthetic esters		PAG	Mineral oil
		TMPTO	Sat/Com		
Biodegrad-ability	Excellent	Very good	Good → Very good	Good	Moderate
Lubrication	Excellent	Very good	Very good	Very good	Good
Stability					
Oxidative	Poor	Moderate	Very good	Good	Very good
Thermal	Moderate	Good	Very good	Good	Good
Hydrolytic	Poor	Moderate	Good	Good	Very good
Viscosity Index	Very good	Very good	Very good	Very good	Moderate
Low temp.	Poor	Good	Good	Good	Good
Seal compat.	Moderate	Moderate	Moderate	Good	Very good
Relative cost	×2	×4	×6−8	×4	1

Key to Table 11.2
Canola Lubricant-grade canola oil containing 60 % w/w oleic acid
TMPTO Trimethylol propane trioleate—a commonly used 'polyol ester'
Sat/Com Fully saturated or complex synthetic ester
PAG Polyalkylene glycol

targets for improved fuel economy, reduced emissions and longer drain intervals have precluded their wider use. A significant uptake of EAL has only really occurred where there has been either a legislative push, or a fiscal pull in the form of financial assistance to facilitate the change from 'conventional' lubricants. The less favorable cost-performance ratio for most EAL has restricted their market acceptance.

As with mineral base oils, the performance of EAL base fluids is improved through additivation. 'Ashless' (metal-free) additives are often used, as heavy metals are normally prohibited. For example, the use of ZDDP (zinc dialkyl dithiophosphate), which is widely used in the industry as an anti-wear/extreme pressure (AW/EP) additive and antioxidant, is not allowed. Other EAL additives tend to be similar to those used in 'conventional' lubricants but with a higher antioxidant treat rate (e.g. base fluids with poor to moderate oxidative stability) and/ or the use of a polymeric thickener/pour point depressant (e.g. canola oil based lubricants). In addition to the usual *performance-cost-treat rate* considerations, the selection of an additive will also be based on its hazards to the environment (e.g. ecotoxicity) or health (e.g. skin sensitization) hazards. Many of the standards for EAL have concentration limits for additives related to these two properties (Battersby 2005) (Table 11.3).

11.2 Environmental Protection Aspects

Water protection: Water legislation stipulates water potential polluting substances. The arithmetic average of three toxicity values (oral mammal toxicity, fish toxicity, and bacterial toxicity) forms the basis of a Water Pollution Number (WGZ) ranging from Water Pollution Numbers 0–6. The highest level of hazard is attained if as little as 1 m^3 of water-miscible cutting fluid is stored. However, a rapidly bio-degradable lubricant remains on the lowest level even if more than 1000 m^3 of the substance exists at the location.

Soil protection: The cleaning of soil is handled in accordance with the list that gives information about the degree of contamination of the soil with mineral oil. It is this degree of contamination that determines whether a clean-up is necessary. The limit for a clean-up is 500 mg mineral oil/m^3 soil. Clean-up costs for soil contaminated with mineral oil can be reduced dramatically if rapidly biodegradable lubricants are used.

Safety at work: The development of rapidly biodegradable lubricants requires the selection of non-toxic materials. Such products are of particular interest in terms of safety at work, and their effect on workplace personnel. Most countries have laws for example, which incorporate numerous guidelines relating to chemicals, categorizes of substances according to their hazard potential and establish classifications.

Toxicity: To develop environmentally acceptable lubricants, toxicological criteria must be considered. The aim is to protect life in various areas, in water (aquatic area) as well as in non-aquatic environments (terrestrial area). The following test procedures are of importance:

> *Bacterial toxicity*: This determines inhibition of cell multiplication. The pseudomonas bacteria type used for this test is found in waste water and soil.
> *Bacterial testing*: This determines acute toxicity through the inhibition of oxygen consumption.
> *Algae toxicity test*: This is a further test for aquatic systems (measurement of chlorophyll fluorescence and determination).
> *Fish toxicity*: Measured the potential pollution of the non-aquatic area, e.g., soil and plants, is evaluated with the plant growth guidelines (for wheat, cress, and rape seed).

Naturally, toxicity testing for environmental protection purposes must also include mammal and human toxicity, as safety at work and environmental protection are increasingly combined. The lethal dose (LDS) is an important measure of toxicity.

Emission thresholds: Every evaporating lubricant pollutes the atmosphere with its emissions, and this is recognized in that, for example engine oil emission limits, and diesel particulate emissions, are regularly reduced. Since the evaporation loss for rapidly biodegradable lubricants is generally lower than that of conventional oils, such lubricants can help meet emission limits. Emissions of flowing metalworking fluids are also often regulated, and several countries have created specific limited threshold values (LTVs).

11.3 Application of Environmentally Friendly Lubricants

Total-loss lubrication: Total-loss applications are the most environmentally sensitive and biodegradable two-stroke oils for boats have been in use since the 1970s. In recent years various special products have also been introduced, and Table 11.4 shows common types of total-loss lubricant available.

    Chain-saw lubricants: This application witness the first imposition of a ban on mineral-oil based chain-saw oils in certain countries with legislation requiring the use of materials that are at least 90 % biodegradable. To encourage the consumption of biodegradable chain-saw oils, some created the environmental lubricant standard to be met.

    Two-stroke engine oils: Two-stroke engine oils are a major part of total-loss lubrication, attention is given here to their use in specific countries. In developing economies two-wheelers are considered more as a personal mode of transport and an essential requirement for everyday use. If this great potential for two-stroke growth, is anything to go by, then reducing the effect of two-stroke engine oils pollution on the environment is a major challenge.

    Concrete mold release: Most mold release oil finds its way into the environment. This means that contamination of soil, water, and air will continue if lubricant suppliers are not able to develop mold-release agents that provide technical performance similar to that of mineral-oil based products. At one point, a biodegradable oil was made but the disadvantage of the product was that it was about three times the price of an equivalent mineral-oil based product.

    Greases: Some of the components traditionally used in greases are inherently biodegradable. As base oils, rape seed oil and many esters are well established in commercial products. Pahi oil, sunflower oil, jojoba oil, and crambe oil remain interesting candidates for a potential use in lubricants. For some time it has been clear that biodegradable greases are good candidates for food-grade lubricants. Several companies have introduced biodegradable greases based on their most popular thickener systems. An interesting approach to this environmental issue has

Table 11.4 Total-loss lubricants currently available

Total loss aspect	Examples
Two-stroke oils	Boating, snowmobiles, and other sensitive areas
Chain-saw oils	Forestry, utilities
Adhesive lubricants	Exposed points on industrial and plant equipment/conveyors
Chain lubricants	Construction, mining offshore
Points grease/flange lubricants	Railways
Concrete mold oils	Construction
Greases	Construction, agriculture, plant equipment
Chain lubricants	Motorcycles, plant conveyors
Corrosion preventives	Agriculture, construction, plant equipment
Wire rope lubricants	Construction, mining, offshore

been the development of an eco-friendly titanium complex grease. It is fairly easy to replace multipurpose greases with biodegradable greases. The technical performance data for ester-based greases are superior to those of any standard multipurpose grease.

Lubrication in railway applications: Another activity that involves lost lubricant during operation is the railway industry. In the past points were often lubricated with used engine oils, hydraulic oils, or greases. There is an increasing trend in wheel flange use of biodegradable grease lubricant. In Canada field tests using bio-degradable rail curve grease have been successfully completed. During these tests it was shown that the bio-degradable grease reduced the coefficient of friction by almost 27 %, halving the wear rate.

Hydraulic oils: Mobile hydraulics has been the focus of attention for a number of years. It can be assumed that a large amount of the fluids sold for such applications ends by polluting the environment. In the field of stationary hydraulics, all applications that have contact with open water or waste-water processing must be taken into consideration. The use of rapidly biodegradable hydraulic fluids in machine tools is a new development, and is justified by a total cost analysis of fluid circuits: a factor influencing this is the cutting fluids used in metal working. The use of ecologically toxic, fire-resistant hydraulic fluids (PCBs) has developed from their use in mining, among other industries. Ester-based fluids (HFDU group) are becoming acceptable in the mining industry. They can display outstanding biodegradability. In terms of volume, water-miscible HFA fluids are perhaps the most important group in mining, where the 1–1.5 % oil phase in emulsions has been replaced by rapidly biodegradable esters. The foodstuff and beverage industries are particularly interested in non-toxic lubricants and fluids, as environmental friendliness and foodstuff compatibility are similar aims.

Engine oils: The influence on future base oil requirements for engine oils would have to focus on improving fuel efficiency (rheology), low evaporation at low viscosity (low emissions) and extended oil change intervals (chemical purity). These specifications will characterize the requirements of base oils in the years to come. The effect of Noack evaporation on diesel particulate emissions is a further reason why limiting the evaporation of oils is desirable for ecological reasons. Lubricant-related particulate emissions account for up to 35 % of total particulate emissions. The dependence of this type of particulate emission on the evaporation characteristics of base oils has been extensively in Fig. 11.1.

Hence there has been a trend to use low viscosities in gasoline and diesel engine oils between 1993 and 2005, resulting in a clear decline in the use of monograde oils, and a move from 15 W-X to low-X oils in diesel engines, and towards 5 W-X oils in gasoline engines as given by tests in Table 11.5.

The increasing performance demands of engine manufacturers are demonstrated by the ongoing reduction in engine oil consumption. On the one hand, this means longer oil-change intervals and on the other hand, falling oil consumption. This oil stress factor is limited not just to oil consumption but also includes

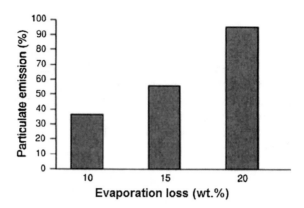

Fig. 11.1 Diesel particulate emissions in relation to evaporation losses of oils (European 13-stage test, CEC R 49, Fuchs Steyr project)

Table 11.5 Emission reduction levels of modern engine oils

	SAE 15 W–40	SAE 5 W–40	SAE 0 W–20
Noak (%)	15.0	9.0	6.9
HC reduction (%)		8.6	26.7
NO$_x$ reduction (%)		17.0	24.5
CO$_2$ reduction (%)		1.9	4.3

continuous specific consumption through exhaust emissions and leakage losses. Both consumption figures have declined considerably over the last few years and this is reflected by higher stress on the oil.

Gear oils: Here the focus is on the development of automatic transmissions. It is interesting to observe that contrary to the situation with engine oil specifications, it can be said that pressure to improve efficiency is leading to engineering changes. Filled-for-life technology will become a more general requirement over the next few years, and conventional gearboxes will be replaced with continuously variable traction units, which make correspondingly greater demands on gear oils. With fill-for-life oils, performance is significantly influenced by the base oils, including the requirements for high oxidative stability, constant friction characteristics, good low-temperature behavior, low evaporation losses and a high natural viscosity index.

Machine tool lubrication: Machine tools are the second-largest consumer of lubricants, after the internal combustion engine. These include varieties of cutting fluids, gear oils, hydraulic oils, slideway oils, spindle oils, greases and other machine lubricants. Cutting fluids are the most important group in terms of volume. Future developments will be determined by three principal factors, safety at work and worker health, the environment, including disposal and system costs. To withstand the considerable potential for contamination by hydraulic, gear, and slideway oils over the life of the cutting fluid, adequate compatibility with machine lubricants must be ensured. Mixing of machine lubricants with the cutting fluids must not cause any deterioration in properties. The correct formulation of additives and base oils in these fluids is essential to prolong cutting fluid life and lower system costs. The contamination of cutting fluids by machine lubricants has been much underestimated. In the case of water-miscible cutting fluids, the tramp oil phase can account for a considerable part of the inner phase, and thus greatly affect the performance characteristics. Every effort should thus be made to separate tramp oils if they are not in stable emulsions. In the case of neat oils, compatible tramp oils can form homogeneous mixtures which have no detrimental effect on the cutting oil.

11.4 Improved Lubricity

Most vegetable oils and synthetic esters display excellent lubricity under boundary lubrication conditions. The high degree of polarity of these lubricants results in their superiority over mineral-oil based lubricants.

Experimental investigations on the twin disc test machines showed that the friction coefficients of vegetable oils and synthetic esters are half those of mineral oils in Fig. 11.2. The experimentally determined friction coefficients were represented as a function of slip, and again, the friction coefficient for the ester is half that of the mineral oil.

11.5 Need for Improved Oxidation Stability

In the diagram below Fig. 11.3 gives the electrical conductivity of various synthetic esters compared to that of rapeseed oil. The electrical conductivity is a measure of the oxidative stability (when oxidation occurs, the electrical conductivity changes). It can be seen that the complex ester has good oxidative stability.

Fig. 11.2 Friction coefficients of base fluids—two disc test $\rho_H = 1000$ N/mm^2; $v_\Sigma = 8$ m/s; $\theta_{oil} = 60$ °C

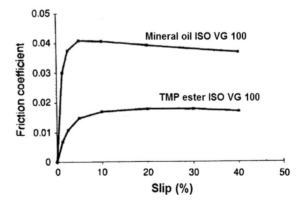

Fig. 11.3 Oxidative stability of biodegradable fluids

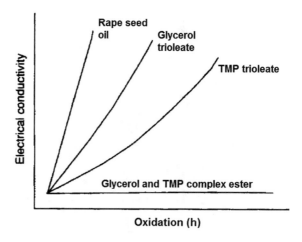

Fig. 11.4 Viscosity-temperature behavior of base fluids (arrows indicate the differences of application temperature between an oil with VI = 100 and an ester based lubricant with VI = 200)

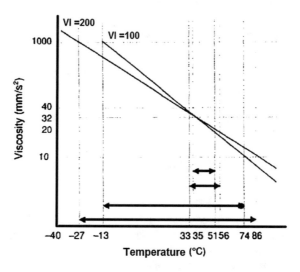

11.6 Viscosity-Temperature Behavior

Using an ester instead of a mineral oil improves the viscosity-temperature behavior of a lubricant. The higher viscosity index of the ester results in a wider temperature range in applications while maintaining the recommended working viscosity, as shown in Fig. 11.4.

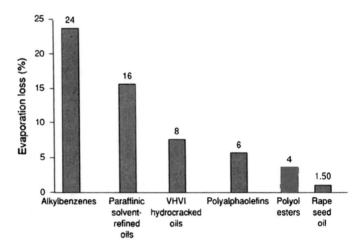

Fig. 11.5 Evaporation loss of base fluids—ISO VG 32

11.7 Evaporation Loss

The evaporation loss of various base fluids has been determined using the Noack test method. As can be seen in Fig. 11.5, the esters and vegetable oils give the best results.

11.8 Bio-lubricant Cost Aspect Advantage

The acceptance of biodegradable lubricants mainly depends on cost. The cost of biodegradable lubricants compared with conventional products is between three and eight times higher. This means that, in terms of their primary costs, such products have to offer considerable additional advantages over traditional materials. Suppliers of such lubricants have the challenge of providing their customers with products which compared to conventional products have not only improved biodegradability but also superior technical performance. To increase quality, including environmental acceptability, in manufacturing, products with greater technical performance are needed, and their benefits must be promoted on the basis of quality rather than cost.

11.9 Current EAL Markets

The current small market share for EAL is typical of niche products sold at relatively high prices but in low volumes. A greater market acceptance will only be achievable if the differences in cost and performance compared with mineral oil based products are reduced. Whilst there have been false dawns in the past concerning the growth in EAL sales, hence a number of initiatives are taken on board. With some EAL based on complex/saturated synthetic esters, the initial high cost of the product may in fact be compensated by a service life that is much longer than a comparable mineral oil based lubricant.

As business in general moves towards a higher level of environmental awareness, *environmentally acceptable products* are developing into *environmentally aware business practices* such as environmental management, product stewardship, life cycle analysis (LCA) and sustainable development. In addition, there are concerns about the security of imported oil and the desire to stimulate rural economies by encouraging the production of 'home-grown' feed stocks for lubricant and energy use. These issues could lead to an increased uptake for EAL based, not only on their biodegradability and eco-toxicity, but also on their sustainability and CO_2 balance (Battersby 2005).

11.10   Bio-based Eco-friendly Lubricants of Recent Times

The emphasis placed on bio-based lubricants is a result of the increase in demand for eco-friendly lubricants that are less toxic to the environment, renewable, and provide feasible and economical alternatives to traditional lubricants. Currently, the interest surrounding liquid lubricants derived from various bio-based feed stocks is focused on the use of eco-friendly lubricants derived from plant-based oils. This is the result of the chemical composition consisting of triacylglycerol molecules made up of esters derived from glycerol and long chains of polar fatty acids. It is these fatty acids that are desirable in boundary lubrication for their ability to adhere to metallic surfaces due to their polar carboxyl group, remain closely packed, and create a monolayer film that is effective at reducing friction and wear by minimizing the asperity contact. Much of the work with eco-friendly lubricants has concentrated on understanding the fundamentals of saturated and unsaturated fatty acids with the bulk of the attention focusing on the use of natural oils as neat lubricants, fatty acids as additives in mineral oils, and bio-based feedstock for chemically modified lubricants. Recently, eco-friendly lubricants are finding uses as carrier fluids for lamellar powder additives in sliding contact (Reeves and Menezes 2016).

Eco-friendly lubricants composed of environmentally benign lamellar powders such as boric acid (H_3BO_3) and hexagonal boron nitride (hBN) are well-known solid lubricants for their low interlayer friction, ability to form protective boundary layers, and accommodate relative surface velocities. As with many lamellar powders, atoms on the same plane form layers through strong covalent bonds. These layers themselves are held together through the weak van der Waals force, providing the minimal shear resistance, and enabling the low interlayer friction. Lamellar powders are effective in a broad range of environments of extreme pressure and temperature as well as various applications from automotive to aerospace to lower friction and minimize wear.

An important property of boron-derived lamellar powders is that they are environmentally benign and inert to most chemicals making them attractive performance enhancing additives to bio-based oils. Experiments have shown that these lamellar particles can be forced out of the contact zone in sliding contact and therefore adding them to natural oils such as canola oil creates a superior eco-friendly lubricant. This new class of eco-friendly lubricant maintains the properties of the powder additives to coalesce and fill in the asperity valleys, thereby establishing a thin, smooth, solid lamellar film between the contacting surfaces, thus decreasing the friction coefficient, wear rate, and surface roughness. In addition, these lubricants maintain boundary lubrication characteristics by establishing the fatty acid adsorption film that thwarts metal-to-metal contact.

11.11 Inadequacies of Current Bio-based Eco-friendly Lubricants

The use of lubricants composed of natural plant oils or solid lubricants have their merits; however, they do have their limitations, which have stifled their ability to be widely accepted within the lubrication industry. The drawbacks to these lubricants are summarized below. For natural oils, they suffer from thermal-oxidative instability, high pour points, inconsistent chemical composition, hydrolytic instability, and a severe susceptibility to biological deterioration. For lamellar powders, they suffer from concentration optimization (which affects their price making these lubricants expensive), unwanted abrasive behavior due to particle size and shape, particles can settle out of the colloidal suspension rendering them useless, large particles can block tubes and capillaries within critical engine parts, and they can clog oil filters in circulatory lubrication systems. These shortcomings of traditional eco-friendly lubricants ultimately cause economic issues where the lubricants themselves can become very expensive when modifying their properties for many thousands of potential applications (Reeves and Menezes 2016).

11.12 Future Bio-based Eco-friendly Lubricants

A new type of eco-friendly lubricant, ionic liquids, is beginning to gain attention. Ionic liquids (ILs) were originally a novel class of solvents typically consisting of an organic cation in combination with any of a wide variety of organic or inorganic anions, exhibit a number of unique and useful characteristics, including high thermal stability, low melting point, a broad liquidus range, and negligible vapor pressure thereby particularly minimizing solvent losses due to volatilization (i.e. fugative emissions), have led many to regard ionic liquids as "green solvents". Thus these properties that make ILs useful in application as good as high-performance lubricants.

Considerable attention has been devoted to the utilization of ILs as lubricants. Three main applications have been most extensively explored: the use of ILs as base oils, as additives, and as thin films. When employed as base oils, ILs have been reported to exhibit good tribological performance for steel/steel, steel/copper, steel/aluminum, ceramic/ceramic, and steel/ceramic sliding pairs. The negligible vapor pressure of ILs makes them good candidates for use under vacuum and in spacecraft applications. ILs are also effective as additives to the main lubricant (e.g., mineral oils), because of their tendency to form strong boundary films, that enhance the tribological performance of the base lubricant. Thin-film lubrication employing ILs has been studied by many researchers with the goal of replacing conventional perfluoropolyether (PFPE) lubricants.

Although the chemical structure of the cationic and anionic substituents of an IL can vary greatly, the most commonly studied ILs in tribological processes have been those containing a tetrafluoroborate (BF_4^-) or hexafluorophosphate (PF_6^-) anion, the result of the superior tribological properties that boron- and/or phosphorus-containing compounds often exhibit under the high pressures and elevated temperatures that lubricants can encounter.In general, as the hydrophobicity of the anion increases, both the thermo-oxidative stability and the tribological properties improve. ILs with longer alkyl chains and lower polarity have been reported to have excellent tribological properties from low to high temperature (-30 to $200\,°C$). Other ILs have been studied with the goal of improving their tribological properties include phosphonium and ammonium (Reeves and Menezes 2016).

The appeal of ILs as lubricants becomes even more evident when one considers their many potential advantages over other lubricants including:

- a broad liquid range (low melting and high boiling point);
- negligible vapor pressure;
- non-flammability and non-combustibility;
- superior thermal stability;
- high viscosity;
- miscibility and solubility;
- environmentally benign (nontoxic);
- lamellar-like liquid crystal structure;
- long polar anion-cation molecular chains; and
- economical costs.

Additionally, ionic liquids have a consistent and easily tailorable chemical composition that affords them the ability to provide the level of thermal-oxidative stability and lubricity required for a variety of applications in the aerospace, automotive, manufacturing, and magnetic storage industries.

The consistent chemical composition allows ILs to have physicochemical properties that are readily reproducible as shown in Fig. 11.6. Furthermore, they can be designed to be eco-friendly by selecting both the cationic and anionic constituents to be nontoxic. In many instances, the ILs can be prepared from nonpetroleum resources. Lastly, their capacity to overcome the variety of environmental, cost, and performance challenges faced by conventional lubricants makes them

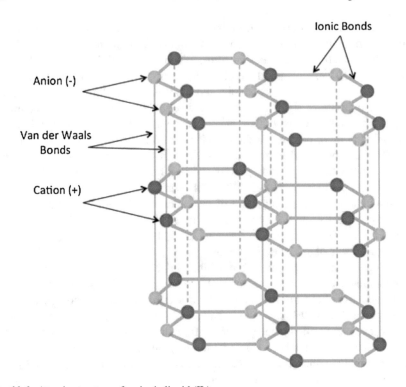

Fig. 11.6 Atomic structure of an ionic liquid (IL)

a potentially attractive alternative eco-friendly lubricant. The possibility of preparing an ionic liquid capable of functioning as an efficient lubricant, while exhibiting a variety of other useful properties is a result of the physicochemical characteristics, inherent tunability, and structural diversity of these novel compounds.

As the market is becoming more ecologically focused with much of the attention centered on novel approaches to achieve efficient energy conservation and sustainability, new classes of green lubricants are being developed that represent the future potential of eco-friendly lubricants. Much of the development aims at creating environmentally friendly lubricants that contain many of the properties of the aforementioned eco-friendly lubricants such as polar molecules (similar to the fatty acids), lamellar crystal structure (like the solid lubricants), derived from bio-based feedstock (natural plant-based oils), high thermal-oxidative stability, physicochemical consistency (which natural oils inherently lack), superior lubricity with minimal wear, and require minimal use of additives. Ionic liquids, particularly those that are fluid at room temperature, represent a promising new class of eco-friendly lubricants that show potential to improve the limitations associated with current petroleum-based oils, bio-based oils, and solid powder additives.

11.13 Conclusion

Biodegradable lubricants offer the best way to minimize pollution in industrial and automotive use. The operation of equipment using high-performance lubricants based on new esters has potential for improvement. Modem types of ester offer outstanding technical properties: their biodegradability and toxicological accept-ability mean they will have an increasing impact on the design of equipment and maintenance procedures. Ester-based products also have economic advantages in terms of safety at work and disposal. These are important cost-saving elements highlighting the need to take into account complete process costs, rather than just the purchase price.

The lubrication industry continues to make new strides toward sustainable eco-friendly lubricants with properties that will lower friction and wear, thereby improving system efficiency and ultimately conserving energy. Continuing this trend, future lubricants formulated from bio-based feedstock should offer the fol-lowing advantages over petroleum-based oils: a higher lubricity lending to lower friction losses and improved efficiency, affording more power output and better economy. As oil prices rise, environmental awareness grows, and the demand for renewable and sustainable lubricants increases, eco-friendly lubricants will begin to seek prominence. Room temperature ionic liquid lubricants represent a new class of novel "greener" lubricants that are nontoxic, obtainable from sustainable (nonpetroleum) resources, and environmentally friendly. They have the potential to satisfy the combination of environmental, health, economic, and performance demands of modern lubricants. Their ability to be tunable establishes them as 'designer' lubricants, where the optimization of cation-anion moiety facilitates energy conservation through superior tribological performance.

Revision Questions

1. *High the common total-loss lubricant applications*
2. *List environmental protection aspects to be considered in lube design*
3. *Outline the role of less viscous engine oils emission reduction*
4. *State advantages of bio lubricants in relation to oxidation stability, coefficient of friction and vaporization loss*
5. *What is the effect of vaporization loss on particulate emission in diesel engine?*

References

Battersby NS (2005) Environmentally acceptable lubricants: current status and future opportunities. In: Proceedings of world tribilogy congress (WTC 2005), Shell Global Solutions International BV, Washington (Sep 12–16, 2005)

Kreivaitis R, Padgurskas J, Spruogis B, Gumbyte M (2011) Investigation of environmentally friendly lubricants. In: The 8th international conference, May 19–20, 2011, Environmental Engineering (selected papers), Vilnius, Lithuania. http://enviro.vgtu.lt

Reeves CJ, Menezes PL (2016) Advancements in eco-friendly lubricants for tribological applications: past, present, and future: ecotribology research developments—materials forming, machining and tribology, vol 2. Springer, Switzerland, pp 41–61. doi:10.1007/978-3-319-24007-7_2, http://www.springer.com/978-3-319-24005-3

Chapter 12
Environmentally Adapted Lubricants

Abstract The present-day requirements for biodegradable and eco-friendly lubricants imply that lubricants have properties that can minimize, if not eliminate, negative environmental impact, such as contamination of soil and water, caused by lost lubrication, leakage and accidents. The dominant, factors that have a direct impact on the environment and which characterize the lubricant and its chemical composition, are toxicity, bio-accumulation and biodegradability. Biodegradability is perhaps the most important factor which determines the fate of lubricant in the environment. Life cycle analysis (LCA) also helps in assessing the total environmental impact of lubricants. This chapter reviews the essential requirements of environmentally adapted lubricants, with regards to chemical composition, eco-toxicity, biodegradability, bio-accumulation, and eco-labeling schemes, and life cycle analysis.

12.1 Introduction

One of the major source of pollution is lubricant as it directly affect the environment when it is handled improperly, as lubricants from petroleum based oils have very poor bio-degradability and also have high toxicity. Environmental pollution can be controlled by the use of biotechnology in the field of alternative lubricants from different available biomass. There is an increasing impact on the environment as a result of lubricants entering the air, soil, and water. It has been estimated that, worldwide, only 50 % of total lubricant consumption is collected, and a considerable amount of the remainder of the 50 % comes into direct contact with the environment due to spillage, machinery failure, leakage throw-off, emissions, and careless disposal. The lubricants lost in this way can endanger the natural environment because of their poor biodegradability and eco-toxic characteristics. Most currently used lubricants are based on mineral oils and chemically derived additives, and harm the environment due to their low biodegradability and toxicity (Jain and Suhane 2013).

© Springer International Publishing Switzerland 2016   165
I. Madanhire and C. Mbohwa, *Mitigating Environmental Impact of Petroleum Lubricants*, DOI 10.1007/978-3-319-31358-0_12

The development of environmentally adapted lubricants is thus necessary to restrict the use of mineral-oil based products in environmentally sensitive areas, such as forests, mines, quarries, railway tracks, agriculture, in water sports equipment, and near drinking water sources. Renewable and synthetic based alternative lubricants have been employed to meet the more severe operating and performance levels in different applications, with proven benefits and advantages over mineral oils.

It is of great concern for all in the oil industry that the environment has been adversely affected due to activities associated with lubricant manufacturers. In the last 10 years, a whole new set of requirements related to the environment has been introduced in different parts of the world, by legislation, through public concern, and through OEM pressures, to deal with the impact of lubricants on the environment. The aim of this chapter is to review some of the environmental issues that relate to lubricants, and to help create awareness of the issues. It also describes the nature of environmentally adapted lubricants (EALs) with regards to their requirements, eco-labeling schemes, and the role of regulations and life-cycle analysis.

In general environmentally friendly lubricants characterizes high biodegradability, low or non toxicity, renewability and exclusion of specific substances. Nevertheless the environmental acceptability of lubricants can also encompass a wide range of potential environmental benefits: resource conservation, pollutant source and emission reduction, recycling and so on. Biodegradability means the tendency of a lubricant to be ingested and metabolized by microorganisms. Complete biodegradability indicates the lubricant has essentially returned to nature. Partial biodegradability bio-degradability usually indicates that one or more components of the lubricant are not degradable. The toxicity and renewability are of grate important too. The fully formulated lubricant shall have the particular carbon content derived from renewable raw materials. One of the most attractive base stocks for environmentally friendly lubricants are vegetable oils. Inherently they have both excellent biodegradability and non toxicity. Moreover vegetable oils have excellent tribological properties, high flashpoints, lower friction coefficient, lower evaporation, and higher viscosity index in comparison with mineral based lubricants (Kreivaitis et al. 2011).

In this chapter, the term environmentally acceptable lubricant (EAL) describes those lubricants that have been demonstrated to meet standards for biodegradability, toxicity and bioaccumulation potential that minimize their likely adverse consequences in the aquatic environment, compared to conventional lubricants. In contrast, lubricants that may be expected to have desirable environmental qualities, but have not been demonstrated to meet these standards, are referred to as environmentally friendly lubricants (EFLs) or bio-lubricants. An environmentally acceptable lubricant should still perform well in comparison to the conventional lubricant it replaces (EPA 2011).

12.2 New Lubricant Requirements

Technical and environmental developments continue to increase the performance requirements of lubricants. The new technical and environmental standards present some of the key challenges facing today's lubricant formulators, in terms of selection of base oils and additive systems. Mineral oils, due to their low biodegradability, are not desirable for environmentally adapted lubricants. The technology is developing with which to produce cheaper hydrocarbons suitable as base oils. These differ from conventional mineral oils from both a technical and an ecological point of view. There is, for example, a growing pressure on lubricant manufacturers to market products that contain recycled oils as an environmental contribution. However, the problem of high polycyclic aromatic hydrocarbons and high chlorine levels in recycled oils makes them unsuitable for producing environmentally adapted lubricants.

   The role of the additive in a lubricant is to enhance the properties of the base oil such that the fully formulated product ensures effective lubrication. Although many of the additives used in mineral-oil based lubricants do not necessarily pose major environmental problems due to their nature and concentration level in the formulated products, it is still necessary to look into the chemistry make-up of the additives more carefully to ensure that they are appropriate for formulating environmentally adapted lubricants, i.e., in terms of biodegradability, toxicity, and bio-accumulation. Governmental regulatory initiatives and 'eco-mark' schemes are also necessary for protecting the interests of consumers. Lubricants have traditionally been developed and assessed purely on technical grounds. Environmental requirements have now compelled lubricant formulators to consider the fate of the lubricant from 'cradle to grave' by life-cycle analysis.

12.3 Base Oil Fluids for Environmentally Adapted Lubricants (EALs)

Because the majority of a lubricant is composed of the base oil, the base oil used in an EAL must be biodegradable. The three most common categories of biodegradable base oils are: vegetable oils, synthetic esters, and polyalkylene glycols. Due to the low toxicities of these three types of base oils, aquatic toxicity exhibited by lubricants formulated from them is typically a consequence of the performance enhancing additives or thickening agents (found in greases) used in the formulation, which can vary widely (EPA 2011).

12.3.1 Vegetables

They are mainly composed of vegetable oils such as triglycerides (natural esters), the precise chemical nature of which is dependent on both the plant species and strain from which the oil is obtained. The most commonly used crops for producing vegetable oil lubricants are rapeseeds canola, soybeans, and sunflowers. And because of performance issues related to low thermo-oxidative stability and poor cold flow behavior, pure vegetable oil-based lubricants comprise a relatively small fraction of the bio-lubricant market, although recent research developments have shown promise for overcoming these shortcomings. Another reason is that vegetable oil-based lubricants are much less available than synthetic esters. To date, their most common commercial applications include hydraulic fluid and wire rope grease.

In addition to their environmental benefits (i.e., high biodegradability and low aquatic toxicity), vegetable oils possess several advantageous performance qualities compared to mineral oils. They have a higher viscosity index (meaning they do not thin as readily at high temperatures) and they have a higher lubricity, or ability to reduce friction. Vegetable oil-based lubricants also have a high flash point, meaning they combust at higher temperatures than conventional mineral oils. They perform well at extreme pressures, and do not react with paints, seals, and varnishes.

Vegetable oils possess several major performance drawbacks, however, which have limited their use in the formulation of EALs. The primary limitations are (1) poor performance at both low and high temperatures and (2) oxidative instability. Vegetable oils thicken more than mineral oils at low temperatures and are subject to increased oxidation at high temperatures, resulting in the need for more frequent oil changes. These shortcomings can be addressed with the use of selected additives for a formulation or through the selective breeding and use of high-oleic oils (i.e., oils that contain more oleic acid, a monounsaturated fat, and less polyunsaturated fats) that are less susceptible to oxidative instability. A lot of chemical modification methods for instance epoxidation, alkylation, radical addition, acylation, hydroformylation, acyloxylation are used to increase oxidation stability. Unfortunately chemical modification decrease inherent biodegradability of vegetable oils making the choice between good technical properties and environmental benefits. Another way to improve above mentioned performance is to use proper additives. However, amount of additives as well as there biodegradability and toxicity is restricted by Eco Labels (Kreivaitis et al. 2011).

The use of selected additives can increase production costs and may decrease the overall environmental acceptability of the product. In addition, vegetable oils remove mineral oil deposits, resulting in the need for more frequent oil filter service. Because the overall formulations are less toxic, disposal costs are generally lower; however, this may not always be the case, as fewer disposal stations are able to accept spent bio-based lubricants.

12.3.2 Synthetic Esters

Lubricants based on synthetic esters have been in production longer than any other class of bio-lubricant and were first used for jet engine lubrication in the 1950s. Synthetic esters can be prepared by the esterification of bio-based materials (i.e., some combination of modified animal fat and vegetable oil). Because synthetic esters can be specifically tailored for their intended application, they have many performance advantages over pure vegetable oils, and are used as the base oil in lubricants for many vessel applications, including hydraulic oil, stern tube oil, thruster oil, gear lubricant, and grease. Synthetic esters-based EALs are developed and marketed by several major oil companies and are currently the most widely commercially available class of EAL.

Synthetic esters perform well across a wide range of temperatures, have a high viscosity index, possess high lubricity, provide corrosion protection, and have high oxidative stability. Because they contain bio-based material, many synthetic esters satisfy testing requirements for biodegradability and aquatic toxicity, although they tend to be less readily biodegradable than pure vegetable oil-based lubricants. Synthetic ester-based lubricants can be more or less toxic than vegetable oil-based lubricants, depending on the aquatic toxicity of the additives used in the formulation. The only notable performance issue with synthetic esters is that they are incompatible with some paints, finishes, and seal materials.

Synthetic esters are generally the most expensive class of EAL. Synthetic ester-based bio-lubricants cost approximately 2–3 times that for comparable conventional mineral oil-based lubricants. As the availability of synthetic ester-based EALs increases, this cost differential is expected to decline. The relatively higher cost of synthetic esters is somewhat mitigated by their high oxidative stability, which results in longer lubricant life. This is particularly applicable to areas of the vessel that require more frequent lubricant changes. Disposal costs are similar to those for vegetable oil-based lubricants.

12.3.3 Polyalkylene Glycols

Polyalkylene glycols (PAG) are synthetic lubricant base oils, typically made by the polymerization of ethylene or propylene oxide. Depending on the precursor, they can be soluble in either oil (propylene oxide) or water (ethylene oxide). Although they are made from petroleum-based materials, PAGs can be highly biodegradable, particularly the water soluble PAGs.

Lubricants consisting of poly-alkylene glycols (PAGs) have the best overall low- and high-temperature viscosity performance among all of the classes of bio-lubricants. The water solubility of ethylene oxide-derived PAGs can improve performance relative to other lubricants by maintaining viscosity following some fraction of water influx (up to 20 % in some laboratory tests), which can be of great importance for stern tube lubrication. PAGs also perform well in terms of lubricity, viscosity index, and corrosion protection. The relatively high viscosity and lubricity of PAGs has resulted in the recent development of PAG-based thruster lubricants.

Disadvantages associated with PAGs are that they are incompatible with mineral oils, as well as most paints, varnishes, and seals. Because of this incompatibility, they have the highest changeover costs of any class of EAL. Additionally, water soluble PAGs may demonstrate increased toxicity to aquatic organisms by directly entering the water column and sediments rather than remaining on the water column surface as a sheen.

12.3.4 Biodegradability

Biodegradability is a measure of the breakdown of a chemical (or a chemical mixture) by micro-organisms. Primary biodegradation is the loss of one or more active groups in a chemical compound that renders the compound inactive with regard to a particular function. Primary biodegradation may result in the conversion of a toxic compound into a less toxic or non-toxic compound. Ultimate biodegradation, also referred to as mineralization, is the process whereby a chemical compound is converted to carbon dioxide, water, and mineral salts. Table 12.1 summarizes biodegradation rates for different lubricant base oils. Ester-based oils have a much greater inherent biodegradation rate due to the presence of carboxylic acid groups that bacteria can readily utilize. These compounds are also more water soluble than compounds that do not contain polar functional groups, the absence of which can reduce their bioaccumulation potential.

12.3.5 Toxicity

The toxicity of lubricants, additives and fuels is the propensity to produce adverse biochemical or physiological effects in living organisms. In addition to possessing a certain percentage of readily biodegradable material, an EAL must also demonstrate low toxicity to aquatic organisms. In general, the vegetable oil and synthetic

Table 12.1 Differential biodegradation rates by lubricant base oils

Lubricant base oil	Base oil source	Biodegradation
Mineral oil	Petroleum	*Persistent/Inherently*
Polyalkylene glycols (PAG)	Petroleum—synthesized hydrocarbon	*Readily*
Synthetic ester	Synthesized from biological sources	*Readily*
Vegetable oils	Naturally occurring vegetable oils	*Readily*

Table 12.2 Comparative toxicity of base oils

Lubricant base oil	Base oil source	Toxicity
Mineral oil	Petroleum	*High*
Polyalkylene glycols (PAG)	Petroleum—synthesized hydrocarbon	*Low*[a]
Synthetic ester	Synthesized from biological sources	*Low*
Vegetable oils	Naturally occurring vegetable oils	*Low*

[a]Solubility may increase the toxicity of some PAGs

ester base oils have a low toxicity towards marine organisms with the LC_{50} for fish toxicity reported as being ~10,000 ppm for fatty acid esters and glycerol esters (see Table 12.2). Water soluble PAGs may demonstrate increased toxicity to aquatic organisms by directly entering the water column and sediments rather than remaining on the water column surface as sheen (Chauhan and Chhibber 2013).

The petroleum-based oils have a greater toxicity to biota in the food chain compared to the other base oil sources. This is related to the more rapid breakdown of petroleum-based oils once in the sea, which ultimately affects the potential for bioaccumulation. The toxicity of petroleum-based oils is also dependent upon additives used in formulations and metabolites produced in biodegradation. The use of additives is dependent on the choice of base oil and the intended function of the lubricant. However, several of the more toxic compounds in formulations are also the ones with poor degradability. The overall product toxicity may be significantly reduced by switching to a biologically sourced base oil used in conjunction with low toxicity additives.

12.3.6 Bio-accumulation

The propensity of a substance to bio-accumulate is another property of a lubricant that is often considered in the qualification of a product as an EAL. Bioaccumulation is the build-up of chemicals within the tissues of an organism over time. The longer the organism is exposed to a chemical and the longer the organism lives, the greater the accumulation of the chemical in the tissues. If the chemical has a slow degradation rate or low depuration rate within an organism, concentrations of that chemical may build-up in the organism's tissues and may eventually lead to adverse biological effects. It is, therefore, desirable to use compounds in formulations that do not bio-accumulate. It may not be possible to phase out all bio-accumulating compounds, but it is feasible to use chemicals that have a lower bioaccumulation potential, either through not being taken up as readily or by degrading more quickly both in the environment and in the organism. The level to which a component of the product is bio-accumulated in an organism is dependent on the environmental and biological half-lives of the compounds (some will degrade before being incorporated into an organism and some will be

Table 12.3 Bioaccumulation potential by base oil types

Lubricant base oil	Base oil source	Potential for bioaccumulation
Mineral oil	Petroleum	*Yes*
Polyalkylene glycols (PAG)	Petroleum—synthesized hydrocarbon	*No*
Synthetic ester	Synthesized from biological sources	*No*
Vegetable oils	Naturally occurring vegetable oils	*No*

metabolized within the organism), as well as the lipophilic nature of the compounds (as measured by water solubility). Any component that has low water solubility may potentially bio-accumulate in an organism. In the case of lubricants, fatty acid-containing components have reduced bioaccumulation potential due to greater water solubility and higher biodegradation rates. This is one distinct advantage in using esters over the other carbon and hydrogen alone base oil types (see Table 12.3).

Currently, a majority of lubricant base oils (mineral oils) have the lowest biodegradation rate, a high potential for bioaccumulation, and a measurable toxicity towards marine organisms. In contrast, the base oils derived from oleo chemicals (vegetable oils and synthetic esters) degrade faster, have a smaller residual, do not bio accumulate appreciably and have a lower toxicity to marine organisms. PAG-based lubricants are also generally biodegradable and do not bio accumulate; however, some PAGs may be more toxic due to their solubility in water. On the basis of this simple comparison, lower environmental impacts will arise if a greater proportion of base oils are manufactured from biologically-sourced materials.

The greater application of environmentally adapted lubricants can significantly reduce the negative environmental impact that has been associated with conventional lubricants. This has become possible by replacing the mineral oils normally used with more eco-friendly base fluids. These new base fluids may be classified as vegetable oils, such as rapeseed oil, soybean oil, and sunflower oil, animal oils such as tallows, manufactured chemicals such as synthetic esters (di-esters, poly-esters, and phosphate esters), or poly-glycols, such as poly-alkylene glycols, synthetic hydrocarbons (poly-alphaolefins, poly-butenes, alkylated benzenes), silicone fluids (poly di-methyl siloxanes, fluoropolymers), and some unconventional HVI mineral oils.

In terms of techno-economic considerations, the cost of vegetable oils today may be two to three times greater than the cost of comparable mineral oils, for the same application, and synthetic esters as much as five to six times costlier than mineral oils. Vegetable-oil based environmentally adapted lubricants have been employed in applications such as conventional two-stroke engines, chain-saws, hydraulic systems, metalworking processes, transformers, concrete-mould release agents, agriculture tractors, and refrigeration systems. Ester-based EALs have been used in two-stroke engines, fire-resistant hydraulic fluids, and metalworking operations, while glycol-based EALs have applications in brake fluid systems, cutting operations, fire-resistant hydraulic systems, gears, compressors, textile fibre processing, rubber processing, and in grease formulation. However, the benefit of using environmentally preferable lubricants can be considerable in terms of reduced environmental impacts.

12.4 Expectation on Additives

Selection of the appropriate additive system is largely dependent on performance specifications, which are driven by (Original Equipment Manufacturers) OEM requirements. However, environmentally adapted lubricants have also to meet environmentally driven criteria, and additional requirements based on legislative, and health and safety considerations, as well as the market desire for enhanced benefits. The majority of lubricant additives consist of long-chain organic compounds to ensure solubility in base fluids. These may generally be poor in biodegradability. They possess low aquatic and mammalian toxicity, and are considered not to pose significant environmental hazards. The main classes of this kind of lubricant additive are succinimides, ashless dispersants, metallic sulphonates, phenates, and salicylates as detergents, zinc dialkyl dithiophosphates as multifunctional anti-wear, anti-oxidant, and corrosion inhibitors. The toxicity data of some of these additives have been reported, and in which case formulators should get a substitute additive.

12.5 Toxicity Levels

Environmentally adapted lubricants should have both biodegradability and good ecological-toxicological characteristics. They should help in protecting life in atmospheric, aquatic, and terrestrial domains. Procedures have been developed in recent years to test the toxicity of substances for aquatic organisms, such as fish, *daphnia magna straus* (a small crustacean), and algae. The best known toxicity classification for impact on aquatic environment is the German based water endangering classification. The class assigned to a substance depends on its WKZ or water endangering number (WEN), which is obtained from measurement of the mammalian fish and bacterial toxicity. The water hazard or water polluting number is determined from the overall WEN, calculated as the average of acute oral mammalian toxicity (AOMT), acute fish toxicity (AFT), and acute bacterial toxicity (ABT) values. The different water hazard classes are defined in Table 12.4.

Aquatic toxicity levels are based on EC_{50} value, which is the concentration of the material which would affect 50 % of the aquatic organisms in a test. LD_{50} (lethal dose) and LC_{50} (lethal concentration) of the material, i.e., which would be lethal to 50 % of the test fish or animals during the exposure period, are other

Table 12.4 Water hazard classification

WEN	WGK	Classification
0–1.9	0	Not hazardous to water
2–3.9	1	Slightly hazardous
7–5.9	2	Moderately hazardous
>6	3	Highly hazardous

important measures of toxicity. Other acute tests include an irritation or sensitization test, in which the degree to which a substance may cause a reaction after contact with the eyes or the skin is measured.

12.5.1 Acceptable Bio-degradability

The bio-degradability of lubricant is primarily influence by the main component that is the base oil, which accounts for 70–80 % in engine oils and up to 90 % in industrial lubricants. The bio-degradability of organic compounds depends on their chemical structures. Hence even esters based fluids used for the production of lubricants can differ in biodegradability from one species to another. The chemical composition (structure) of the compounds that form the composition of the base oils under goes changes during lubricant application as a result of exposure to a variety of factors for example temperature, air, metals, humidity, pressure etc. Above changes in chemical structures produces changes in the properties responsible for the behavior of the oil during service and consequently biodegradability.

Biodegradation is the process in which a lubricant is attacked micro-biologically by bacteria, yeasts, moulds, and fungi, such as are found in soil or water, to break it down into simpler chemicals that the organism can then digest. "Although, the exact biochemical routes of hydrocarbon degradation are not fully known, it is clear that oxygen is important for this process." The substances have been classified according to their degree of biodegradability, as 'readily', 'inherently', or 'relatively' biodegradable. The test methods for 'readily' biodegradable substances determine the extent of ultimate biodegradation. Materials not succeeding in this test may be termed as 'inherently' biodegradable, while 'relative' biodegradability of lubricants can be measured by the CEC-L-33-A-94 test developed by the Co-ordinating European Council (CEC). The Organisation for Economic Co-operation and Development (OECD) and the European Union (EU) have approved these test methods for biodegradability. These test have been generally accepted internationally, and have been in use for several years.

Vegetable oils and its esters are well known to be biodegradable. It is possible to formulate the automotive and industrial lube oil from vegetable oil esters for application such as engine oils, two stroke oils, compressor oils, aviation oil, metal working fluids, insulating oils, gear oils, hydraulic oils and refrigeration oils etc. The trends towards use of alternative refrigerants as replacements for ozone-depleting CFCs, improved heat transfer, filled-for-life refrigeration lubricants and maximization of operational efficiency has made polyol esters and diesters attractive replacement for conventional low temperature petroleum based lubricants. The loss of hydraulic fluids used in agricultural machinery, earthmoving machinery, tunneling machinery, snow crawlers, and deck machinery on ships, waste trucks and road cleaners is uncontrollable and contributes severely to soil contamination because of the toxicity of conventionally used petroleum based and synthetic hydrocarbon based fluids and even phosphate esters (Chauhan and Chhibber 2013).

12.6 Bio-accumulation of Lubricants

The bio-accumulation value of environmentally adapted lubricants is an important measure of the constituent substances accumulation in the fatty tissues of animals or fish, and is expressed in terms of the bio-concentration factor (BCF). The BCF is the ratio of concentration of the chemical in the organism to the concentration of the chemical in water. A simple laboratory test for potential bio-accumulation (OECD 107 and OECD 117) reveals the relative solubility of the substance in water and n-octanol. Water and n-octanol are immiscible, and the substance therefore becomes partitioned between the separate aqueous and organic layers. A preference for the organic layer is reflected in a high value of the n-octanol/water partition coefficient (P_{ow}). The BCF has been shown to correlate strongly with log P_{ow}. This indicates that the bio-accumulation of the substances between P_{ow}, values of >3 and <6 are generally regarded as having a low or negligible potential to bio-accumulate. The bio-accumulation property of a lubricant need not be measured if it is readily biodegradable.

12.7 Eco-labeling Schemes and Regulatory Initiatives

Eco-labeling schemes have already been adopted by a number of countries in Europe and elsewhere as per the directives of their respective governments. Eco-labeling is also an EU-wide scheme (arising from EEC regulation No. 880/92) which was launched in July 1993, to encourage the manufacturers of consumer products to produce items which are less damaging to the environment, and to assist consumers in making informed choice. Products qualifying for an eco-mark are identified by an appropriate logo, and the government is usually entitled to a licensing fee on the sales values.

In other countries, they have promulgated an 'ecomark' scheme for lubricating oils based on vegetable and other oil in which the vegetable-oil based lubricants should have 90 % biodegradability, and lubricants other than vegetable oil based, 60 % biodegradability. The above-mentioned biodegradability criterion is to be tested as per OECD test method CEC-L-33-A-94 (21 days). The product shall not contain any toxic metals, such as lead or barium. Antimony is not allowed in concentrations beyond 0.25 % as tested by AAS methods. The product must not contain halogenated products, such as PCBs, PCTs, and PBTs, or nitrites. The product may contain >50 % by volume re-refined/recycled products, which should be reclaimed through an environmentally compatible re-refining process. Any products containing >2000 ppm of halogens and >20 ppm of PCBs are to be treated as hazardous wastes. The product must not have a toxic effect on aquatic organisms EC_{50}/LC_{50} shall not be less than 1.0 gm/l.

Environmental criteria for an eco-mark labeling scheme for lubricants and greases to limit environmental damage by providing information to the consumer to enable them to select lubricants which are of low risk to have nil effect on the environment by virtue of their superior biodegradability and lesser toxicity, as compared to other products fulfilling the same function. The criteria laid down could be adopted on a wholly voluntary basis for lubricants. This scheme does not in any way address the performance specifications of eco-mark lubricants for their intended applications; however, the product should satisfy normal requirements for intended usage and it should not have significantly poorer properties than normal for the product. Thus the criteria of this scheme are to be subjected to review after 3 years from its implementation.

12.8 The Future of Environmentally Friendly Lubricants (EFLs)

Contaminated environment is expensive. Conventional mineral oil based lubricants are extremely harmful for the biosphere when they get into the environment. Due to poor degradability mineral oils remain in the ecosystem for a long time. Even in case of high dilution the effect will be fatal (eco-toxicological effect). Higher amount will be required for elimination of contaminated ecosystem clearly. Eco friendly hydraulic oil, refrigerator oil, gear oil, motor oil, two stroke engine oils, lubricants for food processing and water management and disposal operations and eco-friendly greases for both general purpose and multipurpose should be widely used. Eco friendly biodegradable lubricants has to be immediately introduced in the market to replace the mineral oil and other non-biodegradable products currently in use in these countries to check rampant pollution caused by these lubricants. Edible oils in use in developed nations such as USA and European nations but in developing countries the production of edible oils are not sufficient. In a country like India, there are many plant species whose seeds remain unutilized and underutilized have been tried for biodiesel production. Non-edible oil seeds are the potential feedstock for production of bio lubricant (Chauhan and Chhibber 2013).

The total environmental impact of environmentally adapted lubricants can be assessed through a complete life-cycle analysis (LCA). This includes both base fluid and additives, the use of raw materials, energy, generation of waste, packaging, distribution, product use, and recycling or disposal. The LCA of most lubricants are not yet available. Instead, claims are usually made on a qualitative comparison between an EAL and its mineral oil equivalent.

Many countries, primarily in Europe, encourage the manufacture and consumption of EALs. Examples are through tax exemptions on environmentally acceptable base oils, taxes on mineral oils, subsidies to consumers to cover the price difference between conventional and EALs, or preferential purchasing programs that require a percentage of certain classes of product to be made from renewable resources (EPA 2011).

12.9   Conclusion

Lubricant formulations must not include certain specific substances, including halogenated organic compounds, nitrite compounds, metals or metallic compounds (with the possible exception of sodium-, potassium-, magnesium-, lithium-, aluminum-, and calcium-based soaps).

Environmentally adapted lubricants are those that can be used in environmentally sensitive applications, e.g., in forests, on railway tracks, near drinking water sources. The components of such lubricants, both base fluids and additives, should be assessed for their toxicity, both human and animal/fish, their biodegradability, according to various test methods, and their bio-accumulation. The ecological performance of such environmentally acceptable lubricants can then be classified with a consumer-aware label. A further step, which has yet to be implemented anywhere on a serious level, is the use of life-cycle analysis, in which the total environmental impact of a lubricant, from inception to ultimate fate ('cradle to grave') is assessed.

For all applications where lubricants are likely to enter the soil and water, EAL formulations using vegetable oils, biodegradable synthetic esters or biodegradable poly-alkylene glycols as oil bases instead of mineral oils can offer significantly reduced environmental impacts across all applications. Although their use is increasing, EALs continue to comprise only a small percentage of the total lubricant market.

Revision Questions

1. *What do understand by EALs?*
2. *Why would there be need to replace mineral base lubricants with EAIs?*
3. *What characteristics should EAL additive and base oil have respectively?*
4. *What do you understand by toxicity and bio-degradability in lubricants?*
5. *How does eco-labeling affect the future marketing of lubricants?*

References

Chauhan PS, Chhibber VK (2013) Non-edible oils as potential source for bio lubricant production and future prospects in India: a review. Indian J Appl Res Chem Vol 3(5)

EPA (2011) Environmentally acceptable lubricants, exhaust gas scrubber washwater effluent. United States Environmental Protection Agency, Office of Wastewater Management Washington (DC 20460)

Jain AK, Suhane A (2013) Mini review biotechnology: a way to control environmental pollution by alternative lubricants. Res Biotechnol 4(3):38–42. ISSN:2229-791X, http://www.researchinbiotechnology.com

Kreivaitis R, Padgurskas J, Spruogis B, Gumbyte M (2011) Investigation of environmentally friendly lubricants. In: The 8th international conference, May 19–20, 2011, environmental engineering (selected papers), Vilnius, Lithuania. http://enviro.vgtu.lt

Chapter 13
Proper Lubricants Handling

Abstract Proper storage and handling of lubricants ensures the high quality which is essential in extending the life of machinery through use of a clean and healthy fluid. Many things can happen to the lubricant between bulk delivery and dispensing to the machinery application. This chapter reviews the general recommended practices for lubricants relating to storage, handling and contamination control, thus protecting the environment in the process.

13.1 Introduction

The major challenges in practice are with regards to storage, handling and disposal of lubricants as summarized by Fig. 13.1 which gives the flow chart of lubricants in terms of the resulting outputs which can possibly adversely affect the environment. As was highlighted in Chap. 1, it is the disposal aspect of the used oil which poses a greatest challenge to environmental protection.

Some contaminated or deteriorated lubricants can be reconditioned for use, while others must be degraded to inferior applications, destroyed or otherwise disposed of. In addition, portions of some contaminated products may be salvaged for use. The decision of which course of action to follow depends on such factors as the amount of product involved and its value compared to the cost of reconditioning or salvaging, the type and amount of contaminant present, the degree of deterioration that has occurred, and the effect of the contamination or deterioration on the functional characteristics of the product in the target applications. It is essential to recognize that, all used oils should be collected for controlled disposals. Some products, such as transformer oils and hydraulic oils, can be readily collected from large industrial concerns, regenerated to a recognized standard and returned to the original source. Oil from the automotive sources will include mono and multi-grade crankcase oils from petrol and diesel engines, together with industrial lubricants that have been inadequately segregated (Udonne 2011).

© Springer International Publishing Switzerland 2016
I. Madanhire and C. Mbohwa, *Mitigating Environmental Impact of Petroleum Lubricants*, DOI 10.1007/978-3-319-31358-0_13

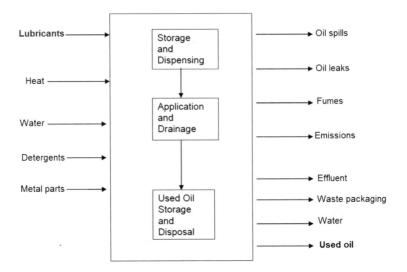

Fig. 13.1 Application and disposal flow chart

The easiest way to control and particularly exclude contamination is to avoid using practices that risk exposing the lubrication to the environment and surfaces/objects that bear various forms and types of contaminants. Among other things, this chapter offers practical advice on how to exclude and monitor contaminants of various types from lubricants during storage and handling. Primary emphasis is placed on the used lubricants since they may contain materials that are harmful to life or the environment or both. Topics of lubricant conservation and the used oil reclamation, reprocessing, and disposal are also addressed and so are the concepts of the environmental compatibility biodegradability and toxicity of the lubricants. Concern for the entry of used lubricant into the environment is on the rise, especially in industrialized countries.

There are three main avenues to restrain the ever-increasing use of lubricants. These are to develop equipment, wherever and whenever possible, that does not require a lubricant, extend service intervals, and when possible recycle the used lubricant. In order to attain the extended service interval, one must use lubricants with extended useful life. Recycling is the option to minimize the used lubricants entry into the environment. This translates into cost savings, with respect to buying a batch of a new lubricant as well as in disposal costs, and the potential damage to the environment, if the disposal method is inappropriate. Ways to minimize inadvertent entry of the lubricant into the environment is to use a closed system, where appropriate.

A prime example is the modern automobile, where the automobile manufacturers have successfully minimized the loss of the lubricant or its volatile components into the environment through leakage and evaporation. They have achieved this by building closely fitting parts and recycling the volatiles into the engine by installing closed ventilation systems. Many industrial users of lubricants employ such self-contained systems to prevent the unintended lubricant loss into the environment.

13.2 Drive for Proper Oil Handling

Environmental regulations governing oil spill prevention, response planning, spill notification and cleanup make little distinction between petroleum oil and vegetable oil. Although used oil waste management regulations exempt vegetable oil products, as a practical matter, many vegetable oil-based lubricants and fluids are not composed of 100 % vegetable oil base stock, but rather are mixtures of petroleum and vegetable oils. In addition, finished vegetable oil-based products contain additives that may consist of regulated substances. Thus storing or managing large quantities of vegetable oil only, mixed petroleum/vegetable oil, or petroleum oil only lubricant or fluid products subjects a facility to essentially the same regulations governing oil spill prevention and response planning.

So any distinction in reporting would come not from what base oil stock is used but rather whether the spilled product contains any listed hazardous substance at levels high enough to require notification. Therefore, proper handling and regulation of greenhouse gas (GHG) emissions, are increasing interest among industrial and commercial users in replacing or reducing the use of petroleum-based products. Vegetable oil-based lubricants and fluids can take advantage of this interest if life-cycle analysis can show reduced GHG impacts compared with petroleum based lubricants and fluids.

13.3 Possible Contamination in Storage

Lubricants degrade for a number of reasons, which may include heat and contamination. To avoid product deterioration, lubricants should only be stored for a limited amount of time as advised by the lubricant supplier.

Solid: Solid contamination includes the additive residue (byproduct of degraded additives), paint chips, rust particles, and weld splatter that may pre-exist within the bulk tank when first commissioned. Solid contaminants can also enter the tank through the breathers, inspection hatches, clean-out portals and through transfer hoses when filling the tank.

Liquid: Moisture, solvents, fuels, and other incompatible lubricants are harmful contaminants as well. Entrained water promotes base oil degradation and additive depletion. Dissolved, emulsified and free water all pose potential risks. In additions to distress imposed by water on additives and base oil oxidation, free water in bulk storage vessels provides a habitat for microbial contamination which is corrosive and harmful to lubricant performance properties. Lubricants in storage are most prone to become contaminated with water from headspace condensation. There are many other sources of water as well. Emulsified water has a tendency to also impair air release properties of oil. When air fails to detrain (release air to the headspace) a common consequence is oil oxidation. Lubricants that are potentially contaminated with volatile products, including diesel fuel, kerosene, or any other solvent, must never be stored in high temperatures.

Thermal degradation: Most good quality synthetic and conventional mineral oils are not affected by storage temperatures below 49 °C. However, storing lubricants near furnaces, steam lines or direct sunlight in high temperature climates for a prolonged time period may cause additives and base oils to oxidize prematurely. A significant darkening of the oil color is an indicator of this condition. Increasing the temperature at which the lubricant is stored by 10 °C doubles the oxidation rate, which cuts the usable life of the oil in half. The presence of water, usually introduced by condensation as a result of temperature variations, increases the rate of oxidation. In greases, the oil may begin to separate from the thickener; this is known as bleeding. The separated oil will typically appear on the surface of the grease, depending on the type of thickener used.

Long-term storage: Some additives in new formulations are not properly dissolved in the oil. When the oil reaches service temperatures these additives may finally dissolve, a process known as "bedding in". Other additives by design will never dissolve. For example, some gear oils may be formulated with solid additive suspensions such as graphite, molybdenum disulfide, as borates. These oils should not be stored for prolonged periods because the solid additives are prone to settle in the storage container.

13.4 Over Fill Protection and Containment

Spills, containment and fire prevention are critical safety and environmental concerns. Depending on the lubricant used, the dangers and environmental impact associated with spills or the likelihood of ignition will vary greatly. For tanks or drums a containment bund wall or material capacity must exceed the volume of lube container by 10 % as shown by Fig. 13.2. A spill catchment basin or dike around the fill pipe that can capture any spilled fluids when the hose is detached can be used to meet this requirement. It does not need to have much capacity because these types of spills are small. Usually a below-grade bucket is used to meet this requirement. Make sure a spill bucket is in place around the fill pipe and verify it is in good condition.

To avoid accidental spillages, overfill protection equipment must be installed. Overfill protection must do one of the following:

- Automatically shut off flow to the tank when it is no more than 95 % full.
- Alert the transfer operator when that tank is no more than 90 % full by restricting flow to the tank or triggering an audible or visual high-level alarm.
- Restrict flow 30 min prior to overfilling, alert the operator with a high-level alarm 1 min before overfilling, or automatically shut off flow into the tank.

It is always advised to place signs on or near the tank should be posted so that personnel are fully aware of the risks associated with the storage tank area. When lubricant is delivered, it is extremely important to be cautious of fire and spill

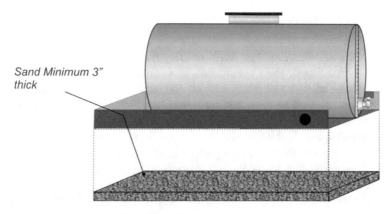

Fig. 13.2 Lubricant spill containment set up

hazards. Plugs and caps should be kept on all hoses and pipes to prevent any dirt or water from contaminating the lubricant at the connection points.

13.5 Used Oil Handling

Lubricating oils, in particular used/waste lubricating oils, represent one of the most hazardous mainstream categories of environmental pollutants. Whether direct or indirect, their short and long term impact on soil, waterways, plants, health of animals and humans is substantial if they are handled or disposed of in an uncontrolled manner. One discharged, their natural degradation can sometimes take up to a few years. Also inadequate burning of used oils, which is a widespread practice, often leads to further significant air emissions of pollutants (Bosina-S Consulting 2006).

Used oil and used oil-contaminated items destined for disposal are subject to requirements for determining whether solid wastes qualify as hazardous wastes. Generators of hazardous waste oil destined for disposal must comply with all hazardous waste management standards. Used oil destined for recycling is subject to assessment to determine if such used oil exceeds thresholds for hazardous waste contaminants and still be managed as used oil, provided the contaminants resulted from normal use of the oil.

Also, assessors must determine whether used oil has been mixed with other materials or waste to determine which regulatory requirements apply. This aspect of the used oil regulations is frequently misunderstood. Mixtures of used oil and ignitable wastes are regulated as used oil if the mixture does not exhibit the characteristic of ignitability. Mixtures of used oil and characteristic hazardous waste are regulated as used oil if the mixture exhibits no hazardous waste characteristic. Used oil is subject to generator, transporter, collection or aggregation point, processor, marketer, and burner requirements.

13.6 Spill Protection

It should be noted that lubricant fluids whether they are petroleum oils, vegetable oils and animal fats or their constituents can cause devastating physical effects, such as coating animals and plants with oil and suffocating them by oxygen depletion. They can as well be toxic and form toxic products. This may result in them destroying future and existing food supplies, breeding animals, and habitats. Also they produce rancid odors, foul shorelines, clog water treatment plants, and catch fire when ignition sources are present. If they are not readily bio-degradable they form products that linger in the environment for many years (Bremmer and Plonsker 2008).

Thus each spill has to be assessed for decisions of action will be based on threats to human or animal populations, contamination of drinking water supplies or sensitive ecosystems, high levels of hazardous substances in soils, weather conditions that may cause migration or release of hazardous substances, the threat of fire or explosion, or other significant factors effecting the health or welfare or the public or the environment. Thus spill guidelines are the same for both mineral lubricants and vegetable based ones.

13.7 Waste Oil Recommended Disposal

A number of used oil regulations allow for four types of recycling which include reconditioning the oil on site to remove impurities; using the oil as a feedstock going into a petroleum refinery; re-refining the oil into a new base stock; or processing and burning the oil for energy recovery. As such proper handling is required before these processes are done to avoid uncontrolled loss into the environment.

Generators, collection centers, transporters, transfer facilities, processors and refiners, and marketers are all key on collection of used oil. In particular, generators of used oil who send their oil for recycling are subject to the following requirements (Bremmer and Plonsker 2008):

- meeting all applicable Spill Prevention, Control and Countermeasures program requirements;
- storing the used oil in tanks, containers, or regulated units (e.g., lagoons, pits, surface impoundments);
- marking "used oil" on the container and tanks;
- keeping tanks and containers in good condition and free from leaks;
- responding to, stopping, and cleaning up releases, and
- having used oil transported by a regulated transporter or meeting self-transporter requirements.

Under current federal regulations, therefore, the waste management requirements for vegetable oil-based lubricant and fluid products require careful review. If a formulated product contains no petroleum or synthetic base oils, it would not be subjected to the used oil management standards. However, the vegetable oil-based product would still need to be assessed after its use to determine if it contains any contaminants that would classify it as a "hazardous waste." Assuming the product contained no such contaminants, it could be managed as a solid waste, providing a potentially broader array of waste management options for the used product (such as burning in small waste heat units).

13.8 Environmental Fate of Lubricants

Used petroleum oils are typically collected and are either re-refined, blended as a supplemental thermal energy feedstock for industrial furnaces, sprayed on coal for dust control or used as rust prevention or friction reduction coatings. Factors influencing the choice for environmental fate include: availability of a facility to re-refine used oil; transportation distance to a re-refining oil facility; availability of an industrial furnace with appropriate air pollution control equipment to handle emissions from used oil combustion; internal uses for used oil; and economics for the life cycle phases of oil production, use and reuse.

Re-refining: When economically feasible, used oils and lubricants are collected and transported to a petroleum oil re-refining facility. Most collection practices do not segregate bio-based oils from petroleum oils and lubes unless the volumes dictate otherwise. If any used oil collection segregation is employed, it is usually for oils that contain emulsified or oil/water mixtures. Thus, it can be expected that used lubes would ultimately be mixed with lubes either at the customer's facility or in the collection truck before going to the re-refining facility. Current re-refining facilities for used petroleum oils are not equipped to separate and recover the bio-oil fraction from a mixture of oils. Petroleum re-refiners are reluctant to accept greater than 2 % bio based lubes in their incoming mixtures due to a concern for degrading the oxidation stability of the base oil they produce. However, there is no data available to substantiate this allegation. The higher boiling fraction would include higher boiling petroleum fractions and be used as an industrial fuel oil feedstock.

Industrial furnace: Where re-refining facilities are too distant for economical transport, used oils are commonly used as a low cost fuel feedstock for industrial furnaces such as aggregate dryers at asphalt plants, cement kilns and blast furnaces. These types of furnaces are fitted with the appropriate kind of air pollution control equipment to limit particulate emissions. These types of furnaces can accommodate a high ratio of soybean oil lubricants in the fuel feedstock.

Other uses: Use lubricants can be use for applications such as dust control for coal and aggregate transported in rail cars and storage piles and as a "once-through" lubricant for oven bearings and chains.

13.9 Lubricant Handling Recommendations at Operational Sites

Lube storage and dispensing: A number of initiatives could be undertaken to ensure that oil is unnecessarily contaminated before it is used. Also there is need to reduce oil leaks and spillages from decanting or malfunctioning dispensers Drums of oils have to be kept under a simple asbestos shed in a head long position to allow for first-in-first out system as drums will be rolled around. This ensures that labels are intact for identification and with readable precautionary instruction in case of emergency. Also drums have to be fitted with suction pumps for drawing out product during dispensing to reduce spills. Where mobility is required a bunded trolley could be provided mounted with a pump to be pushed to all lubrication points to reduce spills. MSDS should be readily available for all product in use for reference in case of emergency or spill. Spill clean-off kits should be also availed to handle spills, this can be in form of absorbent material like wood shavings. In case of bulk storage, fresh oil tanks should be located in bund walled platforms which is 110 % of tank(s) capacity. A comprehensive maintenance system such as Total Maintenance System has to be adopted by workshop management to ensure leak-free tanks and dispensing equipment instead of waiting for the supplier personnel to attend in case of a break down.

Lube application and drainage: The area of lubricant application contributes to environmental pollution in a much more amazing way than is generally thought. This could be due faulty equipment in service which poses major maintenance challenges. It could be decided to start using biodegradable lubricants for remote field sites, which can biodegrade into natural harmless components when they leak. This technology is now readily available for countries like German and Switzerland through Shell Marketing Company. Also adoption of synthetic lubricants is beneficial as these take 2–3 times longer in operation before they deteriorate for drainage as used oil because they withstand high temperature operations even in engine application. This results in reduced oil thrown away. Condition based drainage has to be practiced to ensure that good oil is not thrown away as samples of oil in operation is taken for Tribology tests to determine its condition for continued use. Upon which a decision is made to continue using it or drain. This practice reduces good oil being thrown away into the environment as waste. Mobile collection tank should be availed at field sites to reduce used oil pollution and contamination. Leakage management should be considered on equipment through containment and guttering for leaked oil before it finds its way into the environment. This has to be integrated into Total Productive Maintenance so that potential problems are picked up early before a break down. On draining, proper receivers should be used to collect the used oil from machines in the workshop. Personnel are encouraged to use oil resistant gloves and put on goggles all the time to avoid contact with the skin to avoid cancer. Oil audit to be done on critical equipment to assess what has been dispensed into the machine and what has been drained out monthly basis and take corrective action.

Used oil storage, handling and disposal: All indications are that used oil should be handled with care and properly disposed off in an environmentally friendly manner. All the spills of used oil should be avoided. Hence the following should be implemented in the service workshops:

- Collection from the used oil receivers must be done through as closed system to the storage tank. A pneumatic diaphragm pump should be installed to suck from receiver pan to the tank at the end of each day.
- Used oil collectors to be collected by certified vendors and oil companies who in turn will dispose it by burning it in fire foundries and cement kilns at very high temperature. In both cases the used oil is mixed with paraffin or diesel and used as fuel to fire foundries and kilns. The emission gases are controlled by scrubbers to prevent harmful gases into the atmosphere. While the resulting ash is properly incinerated by the Local Authority Waste Management Department.
- The uptake of used oil from storage tanks should be done using a bulk truck with the required pump and couplings to reduce uptake frequency and incidence of spills.

13.10 Conclusion

Based on economics and ease of handling, the most common "end-of-life" option for used lubricant products is to utilize them as a supplemental fuel source for industrial furnaces. This option offers the widest availability of beneficial users and is competitive with used petroleum based lubricant environmental fate options.

Research has shown that the increased pollution incidents in the environment are more widespread than pollution with crude oil. In recognition of the danger of environmental pollution caused by the indiscriminate disposal of waste oil to individuals and nations, management of waste oil then became a critical cause of concern to nations of the world (Udonne and Onwuma 2014).

Revision Questions

1. Why is there a need for proper handling of used oil?
2. Give reason used oil is considered more hazardous to humans than fresh oil.
3. Name possible contaminants of fresh oil.
4. What is an oil spill containment bund wall?
5. What are the recommended end of life uses of used oil?

References

Bosina-S Consulting (2006) Background analysis for development and establishment of a lubricating oil management system. UNEP/SBC, Sarajevo

Bremmer BJ, Plonsker L (2008) Bio-based lubricants: a market opportunity study update, United Soyabean Board, Omni Tech International, Ltd

Udonne JD (2011) A comparative study of recycling of used lubrication oils using distillation, acid and activated charcoal with clay methods. J Pet Gas Eng 2(2):12–19, Academic Journals. http://www.academicjournals.org/JPGE

Udonne JD, Onwuma HD (2014) A study of the effect of waste lubricating on the physical/chemical properties of soil and the possible remedies. J Pet Gas Eng 5(1):9–14, Academic Journals. http://www.academicjournals.org/JPGE

Chapter 14
Lubricating Grease Handling and Waste Management

Abstract Lubricating grease just like any other lubricants, its waste require careful disposal as it contain pollutants. In response to economic considerations and environmental protection, there is a growing trend of regeneration and reuse of waste lubricants. Accordingly, this work provides an overview on various ways of handling, disposal, treatment of waste grease and its associated environmental impacts. In addition to the fact that petroleum and crude oil are not inexhaustible resources, waste products from these resources present a hazard to human health and the environment. Thus proper management of waste grease is necessary to prevent the adverse environmental impacts, in this regard efficient recycling of waste lubricating greases could help reduce environmental pollution. A review is also made to get an insight into existing literature and what ideas are available for future development through general recommendations regarding the use of renewable resource-based ingredients as a replacement for traditional synthetic materials in conventional greases.

14.1 Introduction

Lubricating greases are made from petroleum lubricants (base oil and additives) which are thickened with metal soaps. Lithium soap is the most widely used thickening agent. Just like in liquid lubricants, some additives are required to enhance performance properties. Thus, the use of polymers is a common practice to modify the rheological properties of greases by reinforcing the role of the thickening agent. The thickener is added to prevent loss of lubricant under operation conditions. Generally, lubrication is achieved by the lubricating oil. Oil viscosity decreases with increasing temperature hence the requirement for additives to support its duties at higher temperatures.

Greases are used in many applications to reduce the wear and friction between movable parts, usually metal joints. Due to its semi solid character, it can stay in place or "put" and act as a seal, thus preventing both solid and liquid contaminants from entering the system, without the need for a sump. Due to the exhaustion

© Springer International Publishing Switzerland 2016   189
I. Madanhire and C. Mbohwa, *Mitigating Environmental Impact*
of Petroleum Lubricants, DOI 10.1007/978-3-319-31358-0_14

of chemical additives and the contamination from metallic materials, particulate dirt and grits, other asphaltic substances as well as the lubricant degradation and its deterioration would result after a certain period of operation. Therefore, it becomes physically and chemically unsuitable for further service use and must be replaced. More than 90 % of lubricating oils and greases are mineral oil based, they are non biodegradable and hence they are not environmentally friendly. It is becoming a growing concern to dispose oils and greases more efficiently and safely. This has led industries and governments to find satisfactory solutions that will reduce the contribution of used lubricants to pollution and also recover these valuable hydrocarbon resources.

14.2 Lubricating Grease Structure Composition

Lubricating grease consists of three basic components: base oils, thickeners, and additives. The present study conducts a comprehensive review of the literature on lubricating grease formulations in all spectra, including their formulations, compositions, and bio-degradability from natural and synthetic sources. Then, renewable resource-based formulations are suggested, and insights and future challenges are identified. The base oils of lubricating greases are categorized as mineral oils (naphthenic and paraffinic oils), synthetic hydrocarbons (polyalphaolefins (PAO) and alkylates), other synthetic compounds (esters and polyglycols), or any fluids with other lubricating attributes; they make up 80–97 % of greases.

As an important component of lubricating greases, thickening agents are often called sponge, given that these semi-fluid materials hold lubricants together (base oil and additives agents). Thickening agents range from simple and complex ones (sodium, lithium, calcium, and aluminum) to non-soapy materials (polymers, bentonites, polyureas, sulfonates, and clays). Other complex thickeners have been widely applied because of their excellent loading capabilities and high dropping points for expected loads. However, the nature of additives that are combined with these thickening agents is believed to have a crucial role in the final product because these additives enhance the homogenization of base oils and the particles of thickening agents, the most applicable one being lithium 12-hydroxystearate soap. Additives modify the microstructures and enhance the existing desirable properties of lubricating greases. Greases come in various forms and compositions and are thus used in different ways and in different fields.

14.3 Applications of Greases

Grease applications vary in terms of their capability to reduce wear and frictional impact in moving machine parts and prevent contaminant interference at application regions by trapping contaminants on the surface. As greases are not completely liquid, their ability to respond to plastic flow at applied forces with-out elastic deformation is immensely important in working conditions and in their function as sealants that prevent lubricating losses. The best assessment of their applications lies in their function and effectiveness during use, which is achievable because of many factors.

Given that the function of greases is enhanced by the materials that constitute them, their adverse environmental impact, the selection of their special func-tions, and their safety during and after use are important issues to be considered, especially in today's era of advancements in technology, industry practices, and industry products. These considerations have given rise to the need to focus on renewable resource-based materials for the formulation of greases. This shift is important because the impact of wastes on the environment is often industrially and globally encountered.

14.4 Grease Manufacturing Process

The grease manufacturing process is complex and sensitive, particularly because it involves various thermo-mechanical properties that may affect the final grease product. The various forms and types of thickeners incorporated in base oils to achieve stable products hamper the development of a universal manufacturing process. Current processes are generally divided into the mixture of inorganic or pre-made soap and thickeners and the in situ production of soap thickeners. In the latter process, fatty acid saponification and dehydration is followed by heating to the near phase transition temperature of soap crystallites, cooling for soap crystallization, milling, and additivation. The former process involves the use of simple constituent mixtures.

Prior to the use of synthetic lubricants, natural oils from rapeseed and castor beans were used, but their disadvantages hindered their continuous application. Notwithstanding the materials used so far in grease formulations and the historical views, advanced studies in lubricating grease technology must still be undertaken, especially because of the wide range of desired applications and expected factors. Moreover, the application of such technology to address future needs should be considered in the context of new electric, hybrid, and biotechnologies. Obviously, many topics remain unexplored. Current and future needs call for considerable improvements and innovations in synthetic lubricants, including the use of natural materials that are 100 % safe in lubricating grease formulations.

14.5 Environmental Compatibility of Greases

The biodegradation of the components of lubricating greases is inevitable, and the same is true for the undesirable substances inherent in such materials. For instance, mineral oil-based greases have wide industrial applications, but they contain traces of nitrogen, metals, and other sulfuric compounds that threaten the ecosystem. These conditions highlight the potential replacement of mineral oils with renewable resource based materials.

Performance, environmental compatibility, and cost are other issues that require attention in grease production. Such issues have emphasized the need to consider renewable resource-based materials for greases. Although several attempts toward improvement have been made, such as the blending of di-esters with PAOs and the blending of synthetic di-esters with canola oils to enhance the biodegradability of greases and reduce their environmental impacts, cost considerations have overshadowed the advantages of these efforts. The evaluation is aimed at demonstrating the feasibility of using renewable resource-based ingredients to replace synthetic ones in industrial grease formulations.

14.6 Biodegradability of Greases

Environmental seals, eco-labeling schemes, consumer awareness of unsafe products and practices, and the disregard for the cost of safer ones have sped up the development of eco-friendly products. Tons of lubricants that are wasted annually present a great risk to the environment and the ecosystem. When miniscule quantities of these materials unexpectedly find their way into the water and soil, they may deter tree growth or cause toxicity to aquatic organisms. This condition has strengthened the demand for effective and safer lubricants that could improve industrial processes and machineries and reduce friction.

A comprehensive review of product improvement from an environmental point of view will thus present significant benefits, especially for consumers who are now acquainted with the negative effects of traditional lubricants and have opted for green products. Moreover, a shift to green products is encouraged as a result of strict government regulations legislated by several countries. Although the use of biodegradable lubricants is not innovative, such lubricants have become increasingly important in various applications, such as gears, transmissions, and engines.

The replacement of mineral oil with vegetable-based materials for safer grease formulations has been achieved because of the safety requirements of lubricant industries. Such promising result is believed to reduce environmental pollutions because of the great bio-degradability of vegetable oils if well adopted. Despite their numerous positive attributes, such as low toxicity, naturally multi-graded properties, excellent solvent for additives, and high load-carrying abilities, vegetable oils also pose certain side effects, such as high costs, low-temperature performance, and low oxidative stability. To overcome these side effects, substituting conventional polyureas or metallic soaps with renewable resource-based materials, and replacing mineral oils with suitable vegetable oils for biodegradable greases are inevitable.

Another motivation behind the use of biodegradable lubricating greases is the consideration of life cycle cost, performance, and environmental friendliness, which are important measures to be undertaken for future lubricants. An estimated 10 million tons of hydraulic liquids and lubricants originated from mineral oils are released to the environment, which can be adversely affected because these lubricants remain non-degradable for a hundred years and cover considerable land expanse. Such conditions lead to poor plant growth and other detrimental effects. This situation necessitates a better and greener replacement. Natural materials that are 100 % safe should be used in this regard.

Other major reasons to go green are the advances in technology, human activities for survival, environmental contamination, particularly lands and water bodies, and health considerations, all of which have become important public issues. In the area of lubrication, focus should be directed toward high volumes of unbreakable lubricants, which pose great threats when released into the environment because they directly contaminate soil and water through their waste. Such waste pollutes the air in the form of lubricant haze or volatile suspended lubricants. The environmental implications of these lubricant wastes are not discounted, resulting in an increasing awareness of their adverse impacts on the ecosystem, the need for technological advancements in final product development, and the possible effects of these products when not properly discarded. An in-depth study into biodegradable synthetic lubricants that can be applied in these environmentally sensitive areas is essential to the efforts toward environmental sustainability. Numerous other threats are inherent in the non-degradable compounds of hydrocarbons, nitrogen, and sulfur with minute amounts of metals found in mineral and synthetic-based formulated lubricants applied in many industries. Such threats have become another motivation to use biodegradable lubricants. In addition, communities must strictly adhere to the REACH process, which involves registration, evaluation, authorization, and restrictions on chemical substances.

14.7 Green Lubricating Greases

Replacing traditional thickening agents that are used in lubricating grease formulations (such as non-soaps, phyllosilicates or polyurea compounds, and metallic soaps) with other environmentally friendly materials may lead to other difficulties because of the way their properties function in the formulation and because of their efficiency conditions, especially metallic soaps. The conditions at which these green materials operate in the context of the latest technologies have heightened the demand for lubricating greases. One such effort is the modification of the oxidative stability of vegetable oils with appropriate antioxidants. Similarly, environmentally friendly base stocks have been improved with rapeseed oil for green lubricant formulations. The latest bio-lubricants are prepared from canola oils. Lithium-based ILs have been applied to high performance lubricants and as viscosity modifiers for performance improvement. Nanoparticle oxides and palm oil-based TMP esters have been employed for wear prevention. Meanwhile, tris (pentafluoroethyl) tri-fluorophosphate ILs have been used to reduce friction and prevent wear (Abdulbari et al. 2015).

Notwithstanding all these efforts, natural biodegradable materials should be introduced as substitutes to conventional materials for grease formulation because they are 100 % safe, both as thickening agents and additives. Such materials have great positive effects on the final products, in addition to their numerous advantages, such as biodegradability, thermal resistance, low toxicity, gel-like characteristics, and high efficiency in friction reduction and wear in the moving parts of machines, which are similar to those obtainable with conventional nonrenewable thickening agents. To reduce the adverse effects of petroleum-based products on the ecosystem, environmentally friendly products that are suitably selected as lubricants are necessary.

14.8 Base Oils for Lubricating Greases

Base oils are the main components of lubricating greases and come in the form of conventional mineral oils and synthetic biodegradable ones, with the former being widely used. Over and above demonstrating excellent performance, lubricating greases must be environmentally friendly and biodegradable. These requirements have initiated investigations into the use of eco-friendly base oils with high oleic acidic content derived from conventional mineral oils.

Basically, there are five types of biodegradable base stocks are of great importance to the lubricant industry: highly unsaturated or high oleic vegetable oils (HOVOs), low viscosity PAOs, PAGs, dibasic acid esters, and polyol esters, all of which are discussed in the present work. Some of these base stocks are biodegradable, whereas others are not. These base stocks could also be natural or synthetic. Among these base stocks, HOVOs are considered as the best substitute for mineral-based products because of their biodegradability, renewability, non-toxic nature, and other important attributes.

Generally, three classes of base fluids are considered biodegradable and environmentally friendly: vegetable oils, synthetic lubricants, and synthetic esters. Synthetic lubricants have been used for many years. The same is true for many synthetic esters, which were developed in World War II and used for many years despite being costly to produce. The demand for renewable resource-based materials as base fluids results from the need to improve the application of lubricating greases, consider their environmental impact, and produce sustainable final products. Given the alarming rate of consumption of fats and oils originating from vegetables and animals, the products and practices related to the production of grease materials require eco-friendly properties, which cannot be achieved in petrochemical resources.

14.9 Vegetable or Natural Base Oils

Natural triglycerides are environmentally friendly esters that feature excellent lubricity and have recently overtaken mineral oils with regard to their adoption in practice. The ability of natural triglycerides to function even in severe and harsh working environments, in addition to their other unique characteristics, has made them the preferred base stock. Vegetable oils are highly biodegradable base oils that present no health hazards in the workplace. However, their use is limited because of their poor thermal/oxidative stabilities, low temperature behavior, and tendency to degrade under severe shear stress, temperature, pressure, and environment conditions. Nevertheless, the triacylglycerols in vegetable oils make these materials an effective base stock for biodegradable lubricants. Vegetable oils comprise long fatty acid chains and polar groups, which contribute to their amphilic nature that makes them applicable in both hydrodynamic and boundary bio-lubricants.

Although organic vegetable oils have high flash points, their low volatility and low vapor pressure help eliminate many dangers in the workplace. Most of the disadvantages of these base stocks could be reduced with the application of appropriate antioxidants. Organic vegetable oils can be used to lubricate chain drives, saw mill blades (at low toxicity), slightly loaded gear drives, or medium pressure hydraulic systems that feature an operating temperature of less than 71 °C, high contamination, and low water ingress potential.

14.10 Chemically Modified Natural Base Oils

Natural triglycerides have certain drawbacks that overshadow their numerous advantages. For example, a major component of natural triglycerides is glycerol, which cannot withstand high temperatures because of its easily destructible nature that results from the abundance of beta hydrogen atoms in its molecules. This limitation has prompted researchers to modify their structures to improve their industrial applications. Such structural modification is often achieved with sulfur incorporation and reaction with thiols and epoxidized vegetable oils without any change in their backbone, but with reduced wear and friction coefficients.

As a type of ester, synthetic polyols are environmentally friendly base fluids with various attributes in high-performance lubricants. However, their volatility, viscosity, and cost have restricted their application in eco-friendly formulations. Hence, they are often blended with mineral oils to boost their performance.

Since esters have immense potential, they have thus been modified and improved to boost their performance. One modification involves the replacement of their beta hydrogen atom structure, which often leads to the partial breakage of their molecules that results in the transformation of unsaturated compounds. When polymerized, these compounds result in precipitate particle formation, which in turn increases liquid viscosity. This modification has been successfully achieved through their replacement with other polyhydric alcohols such as pentaerythritol (PE), neopentyl glyco (NPG), and trimethylolpropane (TMP), which are free from beta hydrogen atoms; despite the tendency of these alcohols to break down at high temperatures, they have unique characteristics and undergo a slow thermal break-age. Synthetic esters in lubricants improve the following properties: thermal stability, hydrolytic stability, solvency, lubricity, and biodegradability.

14.11 Biodegradable Synthetic Base Oils (Esters)

Although esters have certain limitations that affect their life span, such limitations can be addressed by introducing antioxidants and improving their oxidative stabilities. Another improvement in the adoption of polyols as biodegradable polyol esters lies in their preparation via esterification or the transesterification of plant oils and branched neopolyols such as PE and TMP. Other esters such as polyol, diol, triol, and tetrol esters have been developed with the use of catalysts, such as those in solid form, p-toluene sulfonic acid, sulfuric acid, Zr, Ti, Sn, carboxylates, alcolates, or chelates, and tetra alkyl titanatealkylene earth metals.

Most of the biodegradable and renewable resource-based oils discussed above have extensive areas of application, including as lubricating oils. However, they are not effective as base oils in grease formulations. The use of these oils in grease formulations, especially the chemically modified ones, requires validation, and their interaction with different thickeners and their resulting properties should be well noted.

14.12 Thickening Agents

Thickeners are added to lubricating greases to modify the rheological properties and consistency of these greases. Thickeners greatly vary, with the most widely used being fatty acid soaps of lithium, aluminum or barium, calcium, and sodium. In addition to these thickening agents, new thickeners such as synthetic polymers, clays, and biopolymers have been the subject of several investigations. Many natural materials have been used as thickening agents in lubricant formulations. Some of these thickening agents are biodegradable, whereas others are not.

14.12.1 Biodegradable Thickeners

Cellulose: Biopolymer additives in grease formulations resist mechanical degradation and thus ensure superior stability. Cellulose is one of the most widely available polymers in nature and could therefore replace many components in regenerated products such as environmentally friendly thickening agents. Another form of thickening agent in grease is the combination of castor oil and base oil with industrial cellulose pulp or its methylated derivative, the types and concentrations of which influence rheological responses, which in turn lead to improvements in linear visco-elastic (small amplitude oscillatory shear, SAOS) functions.

Kraft cellulose pulp/ethyl cellulose (EC) blends used in lubricating grease formulations provide favorable mechanical stability, which decreases by increasing the EC/kraft cellulose pulp ratio; EC is thus widely applied in many other fields. Kraft cellulose pulp/EC blends are also potential thickening agents because of their capability of improving the physical stability of cellulose dispersions in oils. Extensive efforts have been focused on the development of EC because of its potential as a biodegradable thickening agent, especially when derived from industrial grade cellulose pulp. Kraft cellulose pulp ethylation reactions, ethyl/glucose molar ratio, ethylation time, and effects of ethylation have found that the linear visco-elastic functions of EC are similar to those of traditional lithium lubricating greases. Using cellulose pulp ethylations in obtaining gel-like dispersions may be an excellent technique for formulating environmentally friendly lubricating greases. Gelling agents comprising oleogels such as sorbitan monostearate have also been widely used for biodegradable grease formulations. Most oleogels with this type of thickening agents exhibit tribological characteristics and visco-elastic properties similar to those of standard lithium greases. However, their mechanical stability and thermal dependence are not favorable. These weaknesses leave room for improvements, including the use of cellulose derivatives to enhance thermal resistance, as observed in cellulose derivative thickeners such as castor oil blended with methyl and EC grease. The relative elasticity of oleogels is a function of the constituents of the methylated cellulose pulp/EC blend. Regardless of the cellulose pulp derivative used as a thickening agent, gel-like dispersions usually display degradation at temperatures higher than those of standard lubricating greases. However, methyl cellulose-based gel-like dispersions show poor mechanical stability. This condition paves the way for the development of a new potential thickener in the lubricant industry.

Chitin: Natural ingredients as potential environmentally friendly thickeners have been tested. Chitosan, acylated derivatives, and chitin could replace polyurea or traditional metallic soaps as thickening agents. Oleogels are obtained from biopolymers and characterized thermally and chemically. Some of the rheological and thermal behaviors of oleogels have been studied, along with evaluations of some of their lubricant performance characteristics. Results have shown that the frequency of the evolution of their linear viscoelasticity functions is almost the same as that in standard lubricating greases. The thermal stabilities of acylated chitosan-formulated oleogels are lower than those of chitin and chitosan-based oleogels. Chitin and its derivatives are another group of biopolymers that could serve as potential thickeners for biolubricant applications. Chitin is a polysaccharide obtained from crustacean shells, fungi cell walls, and insect cuticles.

Cellulose and chitin constitute the basis of biodegradable greases if combined with a bio-degradable base oil according to the discussion in the section above.

14.12.2 Non Biodegradable Environmentally Friendly Clay Thickeners

Lubricating greases have been formulated with organically modified clays. Test results have shown that the mechanical characteristics and physicochemical properties of such clays are in line with the extreme-pressure (EP) specifications of greases and thus improve EP properties with EP additives of only 0.5 %. Thus, smectitic clays can constitute new materials for formulating high performance and environmentally friendly lubricating greases with good dispersion and gelation mechanisms. Another important type of thickening agent is the organoclay. Organoclays have been widely used as thickening agents and structure-forming materials because of their abilities to swell and form gel in organic media. They are applied in cosmetics, inks, paints, lubricating grease formulations, and drilling fluids. Organoclays as thickening agents in grease formulations yield high performance and exhibit excellent thermal stability. Organoclays are products of smectite-type clays that are organophilically modified. These clays possess cation exchange capacities ≥ 70 meq 100 g^{-1}. The organophilic characteristics of these clays are induced by a cationic exchange between the interlayer of exchangeable inorganic cations on the clay mineral surface and other ions. At low organic density, ultrasonically treated bentonite organoclays demonstrate enhanced viscosity and thus serve as good thixotropic or thickening agents in lubricating greases. Determining how certain additives, such as silicate mineral powder, can be added to lubricating oils or greases to complement the auto-restoration technology (ART). With ART, in which auto-restoration is enhanced through the grease used in the machines, disassembling machines before they go through a self-repairing process when certain parts wear out becomes unnecessary. Additionally, auto-restoration leads to a spontaneous self-repair for metal surfaces that are already worn out in situ, thereby prolonging the service life of machines.

14.12.3 Synthetic Polymeric Thickeners

Various materials act as gelling materials in lubricating greases and form gel-like dispersions because their thermal stabilities, as determined by certain processing variables, affect their gel formation mechanism; milling or mixing intensities may also exert some influence. Shape and soap particle size often result in the latter. Numerous works have modified and improved the viscous and visco-elastic behaviors of lithium lubricating greases with different recycled or virgin polymers. Different polymers act as effective additives in modifying the rheology of lithium lubricating greases, soap entangled microstructure fillers, and so on.

Olefin blends: The potential of polyolefins/oil blends involving virgin and recycled olefins, such as polypropylenes (PPs) and high-density polyethylenes (HDPEs) dispersed in mineral lubricating oil, are suitable in lubricating greases and have big influence on the resulting microstructures and other rheological properties. It was also found out that, recycled/amorphous polypropylene blends were used as thickening agents in mineral oils; the grease formed exhibited a friction coefficient similar to that of traditional lithium lubricating greases.

Polyolefins as additive in lithium lubricating greases: The swelling and gel formation of polyolefins in organic medium, the strength of lubricants as sealants, or the thickness of films in lubricated contact may expand the industry application of polyolefins as thickeners or additives. Olefins have been used as thickening agents, whereas polyolefins have not been fully explored; although the optimization of the processing protocols needed to form these dispersions has not been well attended to, several studies on the characteristics of polyolefins/oil blends and their numerous applications have been published.

As technologies advance, the need for better-performing greases in terms of operating conditions is also applicable to lubricating greases. Environmentally friendly thickening agents with unique properties may occupy a broad niche in the market place. In this case, cheap and biodegradable recycled polyolefin/oil blends can serve as excellent alternatives. Existing studies have also investigated the effects of adding various polyolefins in minute quantities on the rheological behavior of standard lithium lubricating greases. Polyolefins of different types and molecular weights have been used to formulate greases with consideration of vinyl acetate (VAc) content, rheological characteristics, molecular weight, types and influence of polymers, and crystallinity degree. Generally, the addition of polymers to lithium lubricating greases significantly increases their consistency values. According to, recycled low density polyethylene (LDPE) could potentially be applied as an active rheology modifier in lithium lubricating greases.

When polymers such as styrene-butadiene copolymer, ethylene-propylene copolymers, styrene-isoprene block-copolymers, and poly-isobutylene are incorporated into grease formulations, they demonstrate significant effects as additives, which help to modify the performance attributes of these formulations, including their appearance, tackiness, dropping point, water resistance, and bleed. This condition highlights the suitable rheology properties of polymers as modifiers. The appropriate selection of polymers as additives may be of significant advantage; hence, the right proportion of thickeners, polymers, and base oil should give rise to an optimized formulation with maximum performance characteristics.

However, most of the poly-olefins described in this section do not serve as thickening agents but as additives to modify the rheological behavior of lubricating greases. Although stated that the frictional coefficient of lubricating greases is unaffected by poly-olefins, poly-olefins derived from tribological contact remain the best choice from an environmental point of view because of their low cost, good recyclability, great processability, and excellent physical and chemical combinations.

14.13 Additives in Lubricating Greases

Additives are added to lubricating greases to improve their specific desired properties and characteristics, limit the actions of certain undesired properties, and add new properties, such as mechanical efficiency, friction and wear reduction, and surface damage avoidance under increasing load; these properties can be improved with the addition of the appropriate proportions of additives to lubricating greases. Anti-oxidants remain the most common additives found in greases. Azine and azole derivatives are excellent antioxidant additives because of their anti-corrosion property that prolongs the life span of contact metal surfaces. Another importance of these additives is their anti-wear and extreme pressure property that prevents the excessive wear of metal contact; the addition of both additives to grease yields lubricant-based fluids.

Although the awareness of environmentally friendly materials has led to the introduction of nano-powders such as ceramics, the lack of understanding of the frictional pressure behaviors of these materials has hindered their general acceptance. Thus, talc in the form of suspended ceramic powders has been found to be an efficient additive in environmentally friendly lubricating greases.

The performance of lubricating greases can be influenced by the introduction of a number of additives or a combination of such, the properties of which include extreme pressure, antifriction, anti-wear, anti-oxidants, and so on. As a result of the need for lubricants and other requirements, materials based on nano-particles have emerged in lubricating industries. When these materials are introduced, surface area/volume ratio may increase, which in turn leads to great interaction among tribo contact points and nanoparticles. The use of nanoparticles has numerous advantages, especially in terms of antiwear/friction reduction.

Graphite is another material that displays high lubricity, although it depends greatly on the gases adsorbed. The performance of graphite varies with the environment where they are employed. Carbon particles, a few nano-metals, and nano-diamonds are some of the other materials that feature noticeable lubricating properties, especially when their sizes are reduced or increased in surface areas, but do not exhibit lubricity in bulk or micro scales. Recently, ceramic nano-powders have been tested as anti-wear and EP additives in the form of nitrides, oxides, and sulfides. Other nano-particles have been employed in lubricating grease formulations as additives for property modification.

14.14 Lubricating Grease Challenges

The strict laws related to the use of conventional mineral oils as base stocks in lubricants are the result of their toxic wastes and non-biodegradability. Nevertheless, these conventional materials remain cheaper than biodegradable vegetable based materials despite their inferior properties. As both types of materials have their inherent advantages and disadvantages, a standard measure should be adopted to assist consumers in their product selection process and in enhancing their awareness of environmental considerations, functions, and costs.

Another major challenge is the quantification of the effectiveness of thickening agents. Most of the natural thickeners that are added to lubricating greases have varying effects on grease according to the studies on the contributions to the mechanical stability of the following properties of renewable resource-based materials: rheology modifiers, frictional coefficients, apparent viscosity/visco-elastic function, anti-wear, thermal stability, EP properties, better oil separation, wheel bearing, good tribological response, better lubricity, and self-repair.

Thus, formulations should be made with the right percentage of components to meet certain desired characteristics, and the function of a particular additive should be prioritized for effectiveness. The right selection of ingredients and the resulting microstructures will determine the effectiveness of greases with respect to cost, function, and biodegradability. The last challenge involves the additives used in greases. Given that the major function of greases in lubricating media is to reduce wear and frictional impacts, serve as sealant and offer resistance to plastic deformations, and achieve desired loads, the effectiveness of the additives should be considered. Can greases formulated with certain additives perform their major function and meet other considerations, such as cost, performance, and environmental concerns? Most synthetic lubricants compositions can be altered; therefore, the right selection of materials in terms of functions and loads is expedient. Special care is required when these greases are formulated because the best greases are measured in terms of the function of their components and the resulting microstructures.

14.15 Used or Waste Lubricating Greases

Used lubricants often pose serious pollution problems because of illegal dumping or improper disposal resulting in groundwater, surface water and soil contamination. Direct combustion of these used lubricants without any pretreatment is subject to environmental restrictions as these waste oils may contain toxic and hazardous residues such as metal and metalloid particles, chlorinated compounds, polycyclic aromatic hydrocarbons and other residues which may be released into the atmosphere. Thus, a proper collection system and treatment process for utilizing used lubricants would reduce its environmental impacts as well as preserve valuable resources. 13–15 % of the oil produced is used in open lubricating systems and hence it will unavoidably enter the environment as lost lubricants. Hence the growing drive to promote both use of renewable resources as well as minimizing environmental impact caused by industrial waste materials. Consequently, a new market based on eco-friendly products is developing quite fast, where consumers are determined to use new products or even pay higher prices for alternative materials with reduced negative environmental effects.

The principal source of contamination during grease use is the chemical breakdown of additives and which result in subsequent interaction among the resultant components to produce corrosive acids and other undesirable substances. Among the metals, lead is usually present in high concentrations due to combustion in engines using leaded gasoline. Chlorinated solvents may also be present in significant quantities as a result of the breakdown of additive packaging as well as the presence of chlorine and bromine acting as lead scavengers in leaded gasoline. Polynuclear aromatic hydrocarbons (PAHs) are of particular concern due to their known carcinogenicity. The amount of contaminants in waste oil depends on several factors such as the original detergents and diluents added to the virgin oil.

Although most lubricants are made from mineral oils, some are made from synthetics lubricants, such as silicones. It is important to recover as much as possible of the lubricating oil and recycle it in an environmentally friendly manner. Disposal through incineration contribute to environmental pollution through carcinogenic product emission.

14.16   Impact on Human Species

Exposure to hazardous compounds in waste may be either directly through skin contact or ingestion, or indirectly through the environmental pollution. Health hazards due to drinking waste-oil-contaminated water vary from mild symptoms of accumulation of toxic compounds in the liver to complete impairment of body functions and eventually death. These relate to the hazardous properties of used lubricating oil and the risk of hazardous substances in it. Such impacts may include deleterious effects on the viability and sustainability of terrestrial and aquatic ecosystems.

Indiscriminate dumping of used greases to sewers may contaminate surface waters and interfere with biological treatment and filtration systems at sewage treatment plants. Over time, the layer of grease becomes thicker, and the flow of wastewater in the sewer pipes becomes restricted. This can lead to water and sanitation problems. During heavy rain, sewer overflows can occur and the wastewater in the pipes with plenty of bacteria, pathogens and viruses will result in backup or overflow through manholes into public places, such as streets and parks.

14.17   Impact on Aquatic Species

Used grease in waterways rises to the top, forming a film that blocks sunlight, impairs photosynthesis, and prevents oxygen replenishment, which disrupts the oxygen cycle and enhances growth and reproduction of microorganisms that use oil as a food source. This process leads to eutrophication, whereby available oxygen required for fish, shellfish, and other living organisms which comprise the aquatic food chain is depleted. Furthermore, larval stages of aquatic organisms are particularly vulnerable to toxic substances contained in waste lubricants. Toxins can accumulate in plankton and other tiny organisms at the base of the food chain and ultimately reach human beings as contaminants move up the food chain.

14.18 Economic Impact

When lubricants are disposed in the water system, the layer of grease becomes thicker, and the flow of wastewater in the sewer pipes becomes restricted. Blockages in the wastewater collection system are serious, causing sewage spills, manhole overflows, or sewage backups in homes and businesses. These overflows can result in costly clean-up and repairs as well as attracting severe fines from the regulatory agencies. Countries spend millions of dollars every year unplugging or replacing grease-blocked pipes, repairing pump stations, and cleaning up costly and illegal wastewater spills. These repairs cost money and may lead to higher local wastewater rates, thereby affecting the business.

14.19 Waste Lubricating Grease Management

The purpose of NEMA (National Environmental Management Act 107 of 1998), with regards to waste management, is to avoid waste generation, or to otherwise treat/store/recycle waste in the most sustainable way possible, in order to ensure a healthy environment with reduced impact on human health. The Act states as a principle that waste generation should be avoided, and whenever this is not possible, generation should be minimized. The efficient recycling of waste lubricants could help reduce environmental pollution.

14.19.1 Collection and Handling

Producers of used lubricants have responsibility to dispose their waste in containers which meet appropriate environmental standards. When used lubricant is to be disposed via a third party, the generator should check that the collector is licensed, or otherwise lawfully entitled, to collect and transport such substances. It is also important to specify and record that the collector disposes the waste at an appropriately licensed facility or a facility which is otherwise lawfully entitled to receive it.

14.19.2 Transportation and Storage

Vehicles used for transporting used lubricating oil shall be appropriate for such use, structurally sound and safe. Spillages and contamination of used lubricating oil should be avoided. After its use and application, fresh lubricant becomes waste and it is necessary to temporarily store it at source. Storage tanks and bins in a warehouse or oil house should not be placed near heaters, steam lines or any other plant equipment that generates heat. The next step in the management of waste lubricants is temporary storage at authorized collectors, waste treatment at processors and reporting of data to responsible authorities. However, even though this seems to be a simple procedure, most companies experience lot of problems mainly due to lack of education and separation at source.

14.19.3 Disposal

Majority of lubricants find their way into the environment as many people fail to practice the strict guidelines provided by regulatory authorities. People or companies who pour used lubricants on the ground or into drains or improperly at landfills cause environmental damage. Grease disposal is a challenge; land filling is the conventional and most viable option to dispose waste grease. However, before land filling, water should be removed and the waste grease composition must be known. Some reports have reflected that a large portion of waste end up in landfills and water runoff through accidental leakage. However, this disposal method is a waste of a potential energy source. While attempts for better filtration are increasing, recycling is not a popular option for most companies.

14.19.4 Treatment

Re-processing: The objective of re-processing is to produce a fuel oil with low basic sediment and water content that will not clog burners, foul boiler tubes, or cause sediment build-up in customer tanks. As such, the process requires filtration and removal of coarse solids that can cause environmental or operational problems. Treatment options include mainly physical processes like settling, centrifugation, filtration, or a combination of these operations. Unfortunately, these processes alone are not sufficient to remove all chemical contaminants in the oil, and inclusion of further treatment processes such as clay contacting and distillation would give fuel processors a competitive disadvantage.

Re-refining: Re-fining is the use of distilling or refining processes on used lubrication oil to produce high quality base stock for lubricants or other petroleum products. The use of this method has increased tremendously in developed countries, reaching up to 50 % of some countries' need for lubricating oil. It requires the conversion of waste oil to a product with similar characteristics to those of virgin oil. The re-refining process is a solvent extraction followed by clay treatment or acid treatment. Basically the clay is used as an absorbent. Vacuum distillation followed by clay contacting produces less pollution and it is also an economic solution to the re-refining process, particularly for small scale plants with a capacity range between 10,000 and 30,000 tons. The resulting residual by-product is well compacted and baled in thick plastic sheets prior to disposal in landfills. Participation of a reputable recycling company can play an important role in enhancing the trust factor.

Destruction: This method is preferable in case the waste oil is highly contaminated, particularly with polychlorinated biphenyl (PCB) and polychlorinated terphenyls (PCT). In the absence of hazardous waste incinerators, controlled high-temperature incineration at cement factories is recommended. Temperatures at the flame end of rotating cement kilns ranges between 2000 and 2400 °C. This high temperature is adequate to destroy organics and neutralize acid compounds. The heavy metals content is reduced considerably compared to those found in the natural material used in the cement production process. Continuous monitoring of gas emissions at the cement factories would be required to ensure compliance with air quality standards.

14.20 Conclusion

Although the importance of replacing the conventional materials used in lubricating grease formulations with their synthetic counterparts, which demonstrate better performance characteristics and biodegradability, could not be overemphasized, considering the cost incurred in the use of these materials is equally worthwhile, especially because material costs have an important effect on final products. Natural materials, which offer better biodegradability, safety, and performance, are favorable. Although efforts have been made to combine synthetic and natural materials, especially in a number of applications, which can only be met by distinct grease components, excluding the numerous side effects of these synthetic materials, the cost, performance, environmental issues, and functions required in grease formulation should not be compromised in any form. Therefore, efforts should be made to ensure that natural additives and thickening agents meet the lubricity required for technology development.

Proper waste management provides an opportunity to minimize adverse environmental and health impacts associated with the improper disposal of waste oil. Waste oil is a resource with potential to aid economic growth. Due to high levels of contamination from mineral oil, lubricants such as grease can be based on vegetable or synthetic base oils. Vegetable base oils are biodegradable, renewable and nontoxic.

Revision Questions

1. *What in the basic composition of lubricating greases?*
2. *Give examples of metallic thickeners used in grease manufacturing.*
3. *How can the environmental impact of greases be mitigated?*
4. *Explain re-refining, reprocessing and destruction.*
5. *Unlike liquid lubricant in what way is waste grease difficult to deal with?*

References

Abdulbari HA, Akindoyo EO, Mahmood WK (2015) Renewable resource-based lubricating greases from natural and synthetic sources: insights and future challenges, Weinheim Chem Biol Eng Rev 2(6):406–422, www.ChemBioEngRev.de, WILEY-VCH Verlag GmbH & Co. KGaA

Diphare MJ, Pilusa J, Muzenda E, Mollagee M (2013) A review of waste lubricating grease management. In: 2nd international conference on environment, agriculture and food sciences (ICEAFS'2013), Kuala Lumpur, Malaysia

Chapter 15
Beyond Lubricating Oil and Grease Systems

Abstract This section describes different forms of lubrication, including solid lubricants and how they function. Evidence is provided that anti-seize pastes and anti-friction coatings, fortified with different types and levels of solid lubricants, can provide effective lubrication other than oils and greases in certain applications and support oils and greases in other applications. These pastes and anti-friction coatings (AFCs) can reduce wear, optimize friction and perform under extreme environmental conditions. Also a review on the use of various classes of nanomaterials in lubricant formulations is referred to as having a potential for enhancing certain lubricant properties.

15.1 Introduction

In a basic tribosystem, it is not simply the function of lubricants employed to handle friction and wear, but it also includes opposing parts and their material and surface properties, the loads and relative motions presented by the application under investigation, and the environment under which this entire system is operating (Fig. 15.1).

The properties and characteristics of a typical tribosystem work synergistically to mitigate friction and wear in an effort to improve performance and prolong usefulness of the asset or component within the operating context of the application. Tribological systems are enhanced by the use of various classes of nanomaterials in lubricant formulations. Nanomaterial classes such as: fullerenes, nano-diamonds, ultra-dispersed boric acid and polytetrafluoroethylene (PTFE) are considered. Current advances in using nanomaterials in engine oils, industrial lubricants and greases have contributed to numerous formulation for lubricating materials and indeed nanomaterials do have potential for enhancing specific lubricant properties in future (Zhmud and Pasalskiy 2013).

© Springer International Publishing Switzerland 2016   207
I. Madanhire and C. Mbohwa, *Mitigating Environmental Impact*
of Petroleum Lubricants, DOI 10.1007/978-3-319-31358-0_15

Fig. 15.1 Tribosystem showing parts in relative motion, lubricant film, load and operating environment

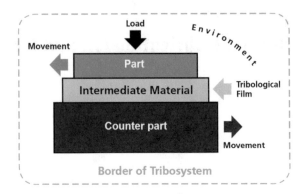

15.2 Lubrication Systems

Generally, film formations are described as hydrodynamic, elasto-hydrodynamic or boundary regimes. A mixed regime can combine elements of boundary and hydrodynamic regimes. Thus lubrication or friction regimes are formed as a function of many different properties and characteristics of the tribosystem. Most important parameters are the speeds, loads, component geometries, substrate material properties and lubricant material properties involved in the application.

Hydrodynamic regime: Fluid lubricants are forced between opposing surfaces as a result of relative motion (speed), and they respond to forces (loads) applied on the opposing component surfaces. The fluid response is to separate one opposing surface from the other with dynamic pressure.

Elasto-hydrodynamic (EHD) regime: In EHD, as loads and speeds increase dynamic pressure, the fluid is compressed to a point at which it begins behaving like a solid. This leads to elastic deformation of the component surfaces, and a transition is made into the EHD regime.

Boundary regime: This is not created by fluid under pressure, but rather by surface-active materials that form boundary films on and between the substrate surfaces. Surface active substances like anti-wear and extreme pressure additives and solid lubricants can adhere to component surfaces as well as cohere to themselves to provide boundary layers. These can protect the substrate from wear by reducing friction. Adhesion and cohesion of surface-active materials are keys to the effectiveness of boundary films.

Mixed regime: This occurs when both hydrodynamic and boundary regimes are present. Under most application conditions, the tribological films will experience each type of lubrication regime at one point or another. For example, during start-up and shutdown periods or in transient events that involve shock loads, conditions can occur in which the relative motion and/or the distance between surfaces converge toward zero. At these points, both partial hydrodynamic films and boundary films are in play.

15.3 Lubrication Concept

Graphically, the liquid film thickness and frictional coefficient can be plotted relative to velocity in (Fig. 15.2). On the left, the load is on surface, while surface and is resting on surface. This marks the boundary regime, or solids contact. As the velocity increases, lubricant is forced between the components in mate. This marks the mixed regime—the boundary regime moving toward the hydrodynamic regime. Dynamic fluid pressure begins to "lift" one component from the other, increasing film thickness between the two. As the film thickness increases, the coefficient of friction decreases.

The point at which the two components no longer touches each other marks the end of boundary lubrication. At this time, the thin film of lubricant between the two components is under enough pressure to deform the surface of the two, and the frictional coefficient is the lowest. This marks the elasto-hydrodynamic (EHD) regime. As velocity continues to increase, the fluid film becomes thicker. As a result, the internal fluid friction of the fluid lubricant and the friction between the two begins raising the frictional coefficient. This marks the full-film hydrodynamic regime.

15.4 Surface-Active Materials

With reference to (Fig. 15.2) at the beginning, the component's velocity is low and the load is extremely high. If the components, are separated by say a coating of wax or with small pebbles. These would be thought of as surface-active

Fig. 15.2 Stribeck curve for friction between surfaces relative to viscosity, speed and load

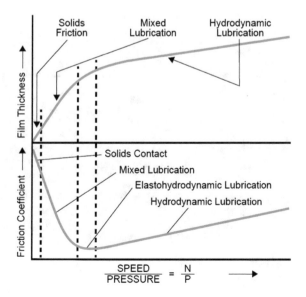

materials to reduce the coefficient of friction when the hydrodynamic film cannot be formed. This is the basic principle behind using solid lubricants to improve lubrication effectiveness and prolong usefulness of assets or components in a given operating application.

15.5 Practical Lubricant Application

In industrial applications, a machine or component requiring lubrication will have a primary lubrication regime based on its steady-state operation. From above each lubrication regime is a function of speed, load, material properties and more. Thus different lubricant forms may be required to provide proper lubrication for the primary regime. In common applications, fluids such as oils and greases, are used to meet requirements of hydrodynamic and elasto-hydrodynamic regimes. Solid lubricants and long polymer-chain additives are used to meet the requirements of boundary regimes. A combination of fluids, additives and solids can be used in the mixed regime.

Therefore, finished lubricant formulations come in many forms to meet different application requirements. Thus for a bearing requiring hydrodynamic lubrication, it will require a lubricant fluid to form the hydrodynamic film. While, a low-speed, highly loaded gear set will require a solid lubricant to form adhesive and cohesive boundary layers to protect gear teeth from wear. A component subject to start-stop conditions and shock loads may need a combination of lubricant forms.

As given in (Fig. 15.3), the boundary regime normally requires solid lubricants, pastes and anti-friction coatings. The mixed regime is best handled with

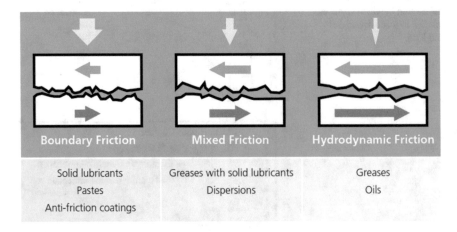

Fig. 15.3 Suitable lubricant forms for different friction states (boundary, mixed and hydrodynamic regimes)

greases and dispersions containing solid lubricants (molybdenum disulphide). And the requirements of the hydrodynamic regime can typically be met with oils and greases. Listed lubricant forms perform best in the specific lubrication (or friction) regimes, yet some cases overlap occurs. This is true for mixed regime in which both the boundary regime and the hydrodynamic regime are present. Thus typical formulation constituent components are found in different lubricants to match different forms work better in different regimes.

15.6   Lubricant Constituent Material

Typical constituents of four common types of lubricant forms are listed below Table 15.1. These include oils, greases, pastes and anti-friction coatings. The base oil (fluid base stock) is the predominant constituent component in oils and greases. This is the primary reason that oils and greases are best for providing hydrodynamic films; lubrication is primarily provided by fluid.

In this regard, solid lubricants are the predominant component in anti-seize pastes and anti-friction coatings (AFCs). This explains why pastes and AFCs are best for providing boundary lubrication films; lubrication is primarily provided by surface-active materials. Note that oils and greases may contain low levels of solid lubricants for supplementary surface wear protection.

AFC solvent: Solvents are commonly found in the anti-friction coating formulations. AFCs use solvents to aid in dispensing and dispersing the solid lubricant and resin components onto the application surface. These solvents evaporate upon application and provide little or nothing to the tribological film formation such is the case in Castrol Grippa 33S for open gears and wire rope application in mines.

Oil and grease: As previously stated, oils and greases rely on fluid to provide the hydrodynamic lubricating films. For these hydrodynamic films to form, relative motion between the opposing surfaces must be present. A key point, however, is that fluid viscosity changes with temperature and pressure, and film thickness changes with viscosity. As a result, in a hydrodynamic regime, the lubricant film thickness will change with temperature and pressure. Also fluids are volatile and subject to issues relating to oxidation, evaporation and the effects of gravity. These

Table 15.1 Lubricant constituent components for specific regime requirements

Constituents in lubricant forms			
Oils	Greases	Anti-seize pastes	Anti-friction coatings (AFCs)
Base oil: ~90 % Additives, including solid lubricants: up to 10 %	Base oil: ~65–95 % Thickener: ~5–35 % Additives, including solid lubricants: ~0–10 %	Base oil: ~40–60 % Solid lubricants: ~40–60 %	Solvents: ~55 % Solid lubricants: ~30 % Resins: ~12 % Additives: ~3 %

inhibit the ability of the fluid to stay where it is needed—between the component surfaces. As a result, lubricant fluids may be unable to form effective tribological films under static loads and high loads, as well as at low speeds (slow relative motion).

Solid lubricants: Unlike the fluids in oils and greases, the solid lubricants that make up a dominant portion of the film-forming functional constituents of anti-seize pastes and anti-friction coatings are relatively unaffected by temperature and pressure. As temperatures and pressures increase or decrease, boundary films formed by solid lubricants can maintain steady thickness without changing as fluids do.

Additionally, because these lubricant materials are in a solid state, they are not subject to evaporation. Oxidation temperatures exceed 399 °C or higher for solid lubricants. Particle size and adhesive and cohesive properties enable solid lubricants to stay in place, even under gravitational influence. The difference between solid lubricants in pastes and AFCs, compared to the fluid lubricants in oils and greases, is that solid lubricants do not rely on relative surface speeds to form tribological films. This is critical under static and high load conditions as well as at slow speeds. In particular, solid lubricants can provide benefits when used in support of fluid lubricants to help protect surfaces during transient events. These can include start-up and shutdown events when surface motion is transitioning from low speed to high speed or from high speed to low speed. The same applies to cases with shock loads brought on by system upsets like pump cavitation or deadheading, compressors being flooded with liquids, bearings under abnormally high vibration, and other such conditions.

15.7 Operation of Solid Lubricants

It was noted that solid lubricants provide advantages in boundary regimes. Understanding how solid lubricants work is important for selecting the most effective lubricant for certain applications. Some of the basic principles of solid lubricants are shown in (Fig. 15.4). Solid lubricants are produced as fine-particle powders. These particles are able to fill in, smooth and cover surface asperity peaks and valleys found on component surfaces. As relative motion and loads are applied to the interacting surfaces, the solid particles adhere to the substrate material. The result is the formation of protective layers to control friction and reduce surface wear.

Because of the particle and surface intermolecular charges, solid lubricants adhere strongly to the surface material and also stick to each other to form lubricating layers. However, solids in powder form are relatively difficult to apply to a surface with much consistency, and keeping them on the surface can be difficult, despite the intermolecular attraction. So, solid lubricants in high quantities normally are applied as a constituent of anti-seize pastes and anti-friction coatings.

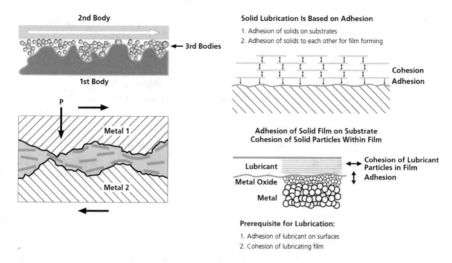

Fig. 15.4   Operating mechanisms of solid lubricants

15.8 Anti-seize Pastes

A paste is a convenient form for easy application of solid lubricants. Typically, such pastes can be used to aid assembly and disassembly of components by providing a consistently lower coefficient of friction than component surfaces alone, even at extreme temperatures. The two primary constituent components in anti-seize pastes are solid lubricants and base oils. Paste formulations typically contain 40–60 % solid lubricants. Various choices for solid lubricants include molybdenum disulfide (MoS_2); graphite; calcium hydroxide; metal phosphates; inorganic oxides; and various metals, like copper, tin, lead, zinc, aluminum and nickel.

The second component in pastes is a base oil, which functions to carry the solid lubricants to the point where lubrication is required. The oil volume will vary between 60 and 40 %, depending on the level of solid lubricants. These carrier oils can be mineral-oil-based or synthetic in nature. Synthetic carriers include polyalphaolefins (PAO), polyalkylene glycols (PAG), diesters (DE), polyolesters (POE), silicones and perfluoropolyethers (PFPE). Paste types are often characterized by color, as this suggests the types of solid lubricants employed in a particular formulation. The different types are summarized in Table 15.2, along with their typical solid-lubricant content and common applications.

Black pastes are mainly composed of MoS_2 and graphite. They are used as assembly pastes to help prevent galling and cold welding and to aid in future disassembly.

White pastes also used as assembly pastes, have metal phosphates and hydroxides as solid lubricants and are particularly good at helping to prevent fretting corrosion caused by micro-vibrations of machine components.

Table 15.2 Types of anti-seize pastes by color, solid-lubricant content and typical applications

Paste type	% Solid lubricants	Typical application
Black pastes	60 % MoS$_2$, graphite and other solid lubricants	Assembly, thin-film
White pastes	60 % white solid lubricants	Assembly, thin-film
Metal pastes	60 % metal powders and other solid lubricants	Threaded connections
Grease pastes	25 % solid lubricants	Thick-film, lifetime lubrication
Oil pastes	10–20 % solid lubricants	Lifetime lubrication, corrosion protection
Metal-free pastes	60 % black and white solid lubricants	Assembly, threaded connections

Metal pastes are often used as anti-seize pastes on threaded connections and can contain various metal powders with other solid lubricants. Some paste formulations may contain small amounts of thickeners, similar to those used in greases, to give some additional consistency and provide "lifetime" lubrication. These are commonly referred to as *grease pastes*, and they typically have about 25 % solid lubricants.

Oil pastes have a reduced amount of solid lubricants—10–20 %, for example—with the balance predominantly oil. These oil pastes are often good for lifetime lubrication and can provide additional corrosion protection and adhesion.

Metal-free pastes are becoming more popular as users seek to become more eco-friendly. These pastes typically use forms of inorganic oxides like zirconium as the solid lubricants. Metal-free pastes are often applied as other pastes are used, but most commonly when high temperatures exist and where solder embrittlement and stress-corrosion cracking can be factors.

15.9 Anti-Friction Coatings (AFCs)

The second form in which solid lubricants can be applied to surfaces is as an anti-friction coating. AFCs are paint-like products in which solid lubricants in a solvent carrier are bound to a surface by a resin material. Like paint, the AFC dries or "cures" to form a thin, dry layer of solid lubricants as the solvent evaporates. The solids provide lubrication for boundary regimes, while the resins as well as the solid-film layers provide some degree of corrosion protection. Typical AFC formulations will contain solid lubricants, resins, additives and solvents. Respectively, solids, resins and additives constitute about 30 %, 12 % and 3 % of the formulation and make up the functional "coating" or film. Solvents make up the balance—about 55 %—of the formulation and serve as agents to aid dispensing and dispersing the solids.

Like pastes, the functional lubricants in an AFC are solids. These solids can include many of the same materials used in pastes, such as molybdenum disulfide (MoS2), graphite and poly-tetra-fluoroethylene (PTFE). Solid lubricant materials, like graphite and MoS_2, typically provide higher load carrying capacity (up to 1000 N/mm^2), while PTFE and other resin waxes provide lower load-carrying capacity (up to 250 N/mm^2) but are typically good at providing a low coefficient of friction in sliding conditions.

The resin or binder system in an AFC provides adhesion of the solid lubricants to the substrate. Resins often have chemical and corrosion resistance properties that complement the surface protection that the solid-lubricant layers provide. In general, the higher the concentration of resin is in the formulation, the better the corrosion protection. Resins can be epoxy, polyamide, phenolic, acrylic or titanate in nature, each offering different cure conditions as well as different adhesion and robustness properties. Organic resins are best for lower application temperatures of 250 °C and below, while inorganic resins are needed for higher application temperatures up to 600 °C.

Solvents help maintain the AFC in a fluid form to aid application and proper substrate coverage. Solvent concentration helps regulate the viscosity during the application process, much like paint thinners can be used to thin paint and promote smooth, even coverage. These solvents may be organic or water-based. Additives play a role in anti-friction coatings, similar to their functional role in oils, greases and pastes. Selected additives can be used to impart additional, desirable properties to the AFC and/or the substrate.

An AFC is typically applied to the surface as a wet film about 30 microns thick. As the solvents evaporate, the resin matrix cures to bind the solid lubricants to the substrate surface in a dry film approximately 15 microns thick. Curing (drying) conditions vary from one AFC formulation to another. This is predominantly controlled by the different resins or binder systems used in the various coating formulations. Cure temperatures can vary from ambient to as high as 250 °C. Cure times also can vary from as short as 5 min to as long as 120 min.

AFCs can be applied effectively in a few ways similar to how paints are applied, such as by spraying, brushing or dipping. Additional methods—like dip-spinning (in which a centrifuge is employed to spin off excess material) or screen printing—can help promote even film thickness and uniform appearance. Just as in painting, AFC effectiveness and service life depend on pretreatment processes. Paint will not adhere to an improperly prepared surface, and neither will an anti-friction coating. Pretreatments, such as degreasing, sandblasting and even acid washing, may be required to remove surface contaminants and allow the resin binder in the AFC to adhere the solid lubricants properly to the substrate. Additional surface treatments, such as anodizing, phosphating or galvanizing, may be considered to help ensure corrosion resistance or provide other desirable properties. AFC resins will still adhere to such pretreated substrates and deliver an extra measure of corrosion protection.

15.10 Advantages and Limitations of Using AFCs

Anti-friction coatings offer many advantages for controlling friction and wear on a surface. Cured AFCs are dry, will not attract dust and dirt, and will work effectively in the presence of dust and dirt. The solid lubricants and resins in AFCs are not susceptible to aging and evaporation like conventional oil and grease lubricants. AFCs can often be used in place of surface treatments like galvanizing to provide both lubrication and corrosion protection. In the event a machine component must sit idle in a prolonged shutdown, AFCs will remain fully effective, unlike oils and greases that may degrade and evaporate. Because solids are used as the primary lubricants, AFCs protect and lubricate across a very wide service temperature range. And AFCs also can be more aesthetically pleasing; when applied correctly, they have a smooth and even surface appearance.

On the other hand, anti-friction coatings are not used as primary lubricants in relatively high-speed, rolling applications. Higher speeds typically require lubricating films for hydrodynamic regimes, and rolling applications typically require lubrication for elastohydrodynamic regimes. These regimes are best served by fluid containing lubricants like oils and greases. Yet, AFCs may be used as secondary, emergency or transient event lubricants to support a primary oil or grease lubricant. An example would be using an AFC to protect a pump shaft at start-up or shutdown, when shaft speeds are such that sleeve bearings are not operating at speeds where hydrodynamic films can be formed. Another downside to AFCs is that the application process can be somewhat costly. Special equipment, extra handling and trained coating shops may be needed. Also, required cure times may not always fit into existing manufacturing processes; components may require coating by other resources. As noted, however, such added costs can be overcome by reducing or eliminating re-lubrication requirements.

15.11 Typical Applications of AFCs

There are countless applications for AFCs. Generally, these include low-speed and high-load applications that require boundary regime lubrication. AFCs also are good for dusty or dirty environments in which oil and grease lubricants can attract contaminants that may lead to accelerated abrasive wear. Applications where oscillatory motion and vibration can cause fretting corrosion also are good candidates for an anti-friction coating. In addition, AFCs can help reduce premature machine wear from initial start-up and run-in operations. They also can provide good corrosion protection, replacing heavy metal coatings to prolong asset life and enhance environmental friendliness. Applications with sliding friction and wear mechanisms—such as cams, slides, ways and springs—also are ideal for AFCs.

15.12 Drive for Solid Lubricants

Recently, however, a number of new applications have arisen that have prompted renewed interest in solid lubrication. These include lightweight moving mechanical assemblies and tribological components for long-term service in space mechanisms, and cages for turbo pump bearings operating in liquid hydrogen and oxygen. The new requirements are primarily long-term life and successful operation over a broad temperature range. New solid lubricants are needed to meet these requirements.

The success of habitats and vehicles on the Moon and Mars, and ultimately, of the human exploration of and permanent human presence on the Moon and Mars, are critically dependent on the correct and reliable operation of many moving mechanical assemblies and tribological components. It is essential, therefore, to have a thorough understanding of tribological components, such as bearings and gears, and of how to select the right lubrication for each application. This may require designing for new solid lubricants and design validation efforts in applications where liquid lubricants are ineffective and undesirable. Environmental interactions will have to be considered carefully in the selection and design of the required durable solid lubricants. Several environmental factors may be hazardous to performance integrity.

The most commonly used solid lubricants and their characteristics are summarized in Table 15.3. Solid lubricants are used when liquid lubricants do not meet the advanced requirements of modern technology. Oils or greases cannot be used in many applications because of the difficulty in applying them, sealing problems, weight, or other factors, such as environmental conditions. Solid lubricants may be preferred to liquid or gas films because they reduce weight and simplify lubrication. For many applications, solid lubricants are less expensive than oil and grease lubrication systems.

Under high vacuums, high temperatures, cryogenic temperatures, radiation, high dust, or corrosive environments, solid lubrication may be the only feasible system (Table 15.4).

In extreme pressure conditions such as high to ultrahigh vacuum conditions— a vacuum of $\sim 10^{-2}$ Pa), such as space, lubricants can volatilize. In high-vacuum environments (such as space-vacuum environments), a liquid lubricant would evaporate and contaminate the device, such as optical and electronic equipment. Also in extremely high temperature conditions, liquid lubricants can decompose or oxidize. Suitable solid lubricants can extend the operating temperatures of systems beyond 250 °C or 300 °C while maintaining relatively low coefficients of friction. In cryogenic temperatures, liquid lubricants can solidify or become highly viscous and not be effective. Suitable solid lubricants can extend the operating temperatures of systems down to −273 °C. In radiation environments, liquid lubricants can decompose. Suitable solid lubricants can extend the operation of systems beyond 10^6 rads while maintaining relatively low coefficients of friction. In high dust areas, hard solid lubricants, such as diamond like carbon and boron nitride film, are useful in areas where liquid lubricants tend to pick up dust.

Table 15.3 Common solid lubricants and characteristics

Solid lubricant	Characteristics
MoS_2	MoS_2 has a low coefficient of friction both, in vacuum and atmosphere, and it does not rely on adsorbed vapors or moisture. Its thermal stability in nonoxidizing environments is acceptable to 1373 K, but in air the temperature limitation of MoS_2 may be reduced to a range of 623–673 K by oxidation. Adsorbed water vapors and oxidizing environments may actually result in a slight, but insignificant, increase in friction. MoS_2 has greater load-carrying capacity than other commonly used lubricants, such, as graphite and PTFE. MoS_2 has a hexagonal crystal structure with the intrinsic property of easy shear. The lubrication performance of MoS_2 often exceeds that of graphite, and MoS_2 is effective in vacuum where graphite is not
Graphite	Graphite has a low coefficient of friction and very high thermal stability (2273 K and above). Graphite has a hexagonal crystal structure with the intrinsic property of easy shear, although graphite relies on adsorbed moisture or water vapors to achieve low friction. Use in dry environments, particularly in vacuum, may be limited. At temperatures as low as 373 K, the amount of water vapor adsorbed may be reduced to the point that low friction cannot be maintained so sufficient water vapor may be deliberately introduced to maintain low friction. Practical application at high temperatures is limited to a range of 773–873 K because of oxidation. When necessary. additives composed of inorganic compounds may be added to enable use at temperatures to 823 K
PTFE[a]	PTFE has a low coefficient of friction both in vacuum and atmosphere because of a lack of chemical reactivity. PTFE does not rely on adsorbed vapors or moisture. It possesses low surface energy and does not have a layered structure. The macromolecules of PTFE slip easily along each other, similar to lamellar structures. Practical application temperatures range from 173 to 523 K. PTFE does not have greater load-carrying capacity and durability than other alternatives. The low thermal conductivity of PTFE inhibits heat dissipation, which causes premature failure due to melting and limits use to low-speed sliding applications where MoS_2 is not satisfactory. PTFE shows one of the smallest coefficients of static and dynamic friction, down to 0.04. Operating temperatures are limited to about 523 K
Soft metals	Lead, gold, silver, copper, indium, and zinc possess relatively low coefficients of friction both in vacuum and atmosphere because of their low shear strengths. These metals are extremely useful for high-temperature applications up to 1273 K and for rolling element applications, such as roller bearings, where sliding is minimal

[a]Polytetrafluoroethylene

These contaminants readily form a grinding paste, causing abrasion and damaging equipment. The elimination of liquid lubricants and their replacement by solid lubricants would reduce spacecraft weight and, therefore, have a dramatic impact on mission extent and craft maneuverability. Under intermittent loading conditions or in corrosive environments, liquid lubricants become contaminated. Changes in critical service and environmental conditions—such as loading, time, contamination, pressure, temperature, and radiation—also affect liquid lubricant efficiency. When equipment is stored or is idle for prolonged periods, solid lubricants provide permanent, satisfactory lubrication. The pros and cons of solid lubricants are given in Table 15.5 below.

Table 15.4 Areas where solid lubricants are required

Requirement	Applications
Resist abrasion in dirt-laden environments	Space vehicles (rovers), Aircraft, Automobiles, Agricultural and mining equipment, Off-road vehicles and equipment Construction equipment, Textile equipment
Avoid contaminating product or environment	Medical and dental equipment, Artificial implants, Food-processing machines, Metalworking equipment, Hard disks and tape recorders, Paper-processing machines, Automobiles
Maintain servicing or lubrication in inaccessible or hard-to-access areas	Space vehicles, Satellites, Aerospace mechanisms, Nuclear reactors, Consumer durables, Aircraft
Provide prolonged storage or stationary service	Aircraft equipment, Railway equipment, Missile components Nuclear reactors, Heavy plants, buildings, and bridges, Furnaces

Table 15.5 Advantages and disadvantages of solid lubricants

Advantages	Disadvantages
• Are highly stable in high-temperature, cryogenic temperature, vacuum, and high-pressure environments • Have high heat dissipation with high thermally conductive lubricants, such as diamond films • Have high resistance to deterioration in high-radiation environments • Have high resistance to abrasion in high-dust environments • Have high resistance to deterioration in reactive environments • Are more effective than fluid lubricants at intermittent loading, high loads, and high speeds • Enable equipment to be lighter and simpler because lubrication distribution systems and seals are not required • Offer a distinct advantage in locations where access for servicing is difficult • Can provide translucent or transparent coatings	• Have higher coefficients of friction and wear than for hydrodynamic lubrication • Have poor heat dissipation with low thermally conductive lubricants, such as polymer-base films • Have poor self-healing properties so that a broken solid film tends to shorten the useful life of the lubricant (However, a solid film, such as a carbon nano tube film, may be readily reapplied to extend the useful life.) • May have undesirable color, such as with graphite and carbon nano tubes

The effectiveness and performance of hard to super hard coatings—such as diamond like carbon, WC/C, TiC/C, VC, diamond films, and other solid-film lubricants—have been validated in a variety of sliding contact conditions in the atmosphere by many researchers in industry, academia, and government. The friction and wear interactions of solid lubricants and coatings are system properties.

This means that performance and behavior depend on the lubricant and materials, the operating conditions, and the system.

Many of the properties of solid lubricants and coatings are actually surface properties. For example, friction, adhesion, bonding, abrasion, wear, erosion, oxidation, corrosion, fatigue, and cracking are all affected by surface properties. By depositing thin films, producing multilayered coatings, and modifying surfaces, designers can enhance performance, that is lower surface energy, adhesion, and friction, and increase resistance to abrasion, wear, erosion, oxidation, corrosion, and cracking, as well as improve compatibility with the lunar and Martian environments. The study already done indicated that thin films and coatings are commonly used in components and devices to improve mechanical properties, material performance, durability, strength, and resistance in basic industries, such as industrial coatings, nanotechnology, optical components, plastics, ceramics, biomedical technology, instrumentation, micro electromechanical systems, and disk drives.

15.13 Nano-materials and Limitations in Lubricant Application

The continuing pursuit for better fuel efficiency stands behind many recent advancements in engine technology. The introduction of higher power densities in modern diesel engines raises performance requirements for engine oil. Hence nano-additives open new ways to maximizing lubricant performance. Even though nano-materials have been around for quite a while, and numerous studies have been carried out showing that nanotechnology can indeed improve the lubrication properties of oils and greases, large-scale market introduction of nano-fortified lubricants is still facing serious technical and legislative obstacles (Zhmud and Pasalskiy 2013):

Fullerenes: Fullerenes are cage molecules which are claimed to enable "rolling" lubrication mechanism. This has never been actually proven. The C60 carbon material has been best studied. Inorganic fullerenes comprise another class of nano-materials with "fullerene" tag. For instance, inorganic fullerene-like material (IF-WS$_2$) nano-particles can be synthesized by reacting sulfur with tungsten trioxide (WO$_3$) nano-particles in a hydrogen atmosphere at 500–650 °C. The IF-WS$_2$ nano-particles have a closed hollow cage structure with an average size of about 50 nm, which is much larger than the size of the C$_{60}$ molecule. Studies suggest that addition of C$_{60}$ fullerene soot in a lubricant significantly increases the weld load and seizure resistance. C$_{60}$ fullerene soot and IF-WS$_2$ nano-particles form much more stable dispersions in hydrocarbons as compared to regular graphite and WS$_2$ powders. IF-WS$_2$ is marketed as the extreme pressure/anti-ware additive for engine oils, gear lubricants and greases, yet its applications so far are very limited. Among the chief limiting factors is the uncertainty about the health safety and environmental (HSE) profile of fullerenes. IF-WS$_2$ also has issues with copper corrosion and poor oxidation stability. Changes in various performance characteristics of a motor oil due to deployment of IF-WS$_2$ in formulation are modest improvements in wear protection and fuel economy, though these are outweighed by degradation in such pivotal properties as corrosion protection, with a specific risk for main bearing corrosion, oxidative thickening, and emission system durability.

Nano-diamonds: The term is usually used to describe ultra-dispersed diamonds produced by detonation of hexagen or trinitrotoluene in a closed chamber. The average particle size is 4 to 6 nm. As a lubricant additive, nano-diamonds are claimed to embed into the sliding surfaces rendering them more resistant to wear, or alternatively, enable "rolling lubrication" between the surfaces, thus reducing friction and wear. The extreme pressure/antiwear efficiency of nano-diamonds was shown to result in a reduction in friction when nano-diamonds are added to lubricant formulations, and is consistent with the micro-polishing effect resulting in faster running-in and smoother mating surfaces. A similar effect has been observed for carbon nano-horns, and as a result the transition from full-film to boundary lubrication occurs at a lower velocity-to-pressure ratio. Since the abrasiveness of nano-diamonds does not go away after the initial running-in period, there is a risk for excessive wear over a longer period of time. Analysis of oils from engines run with engine oils doped by nano-diamonds reveals unusually high levels of wear metals such as aluminum, copper and chromium, indicative of accelerated wear of bearings and piston rings. The micro-polishing effect of nano-diamonds in engine oil seems to improve surface finish of certain components after the running-in. Therefore, nano-diamonds may prove useful in running-in oil formulations, yet more studies are needed to discern possible unintended consequences.

Boric Acid: Boric acid used to be a common additive in metal-working fluid (MWF) formulations due to its excellent EP/AW properties, and bacteriostatic as well as bactericidal actions. Some recent studies showed that boron-based nano-particulate lubrication additives that can drastically lower friction and wear in a wide range of industrial and transportation applications. By replacing sulfur and phosphorous, boron additives are hoped to eliminate the main sources of environmentally hazardous emissions and wastes. Unfortunately, there are quite a few technical hurdles to mar that optimism, as boric acid has no antioxidant effect, so it cannot replace zinc dithiophosphate (ZDDP); and boric acid is not compatible with some essential lubricant additives, specifically with the total base number TBN)buffer in the engine oil, which may lead to corrosion and sludge problems.

Polytetrafluoroethylene (PTFE): PTFE has a well-defined footstep in the lubrication engineering with impressive performance profile in greases, chain oils, dry-film lubricants, etc. Recognition of potential to reduce friction and wear has led to use of PTFE as a dry-film lubricant and friction modifier long before the buzzword "nano" has come into daily use. PTFE-fortified oils and greases are known to exhibit higher welding loads, higher load wear indexes, and reduced stick-slip. Though PTFE nano-dispersions are used in a number of aftermarket engine treatment products, the use of PTFE in engine oils is rather limited because of inherent instability of PTFE dispersions in oil, the risk of oil filter clogging, as well as difficulties with recycling. As a matter of fact, the application of PTFE in engine oils is discouraged even by the major PTFE producer.

Nano-materials have potential for enhancing certain lubricant properties, yet there is a long way to go before balanced formulations are developed.

15.14 Conclusion

Oils and greases can be effective in hydrodynamic regimes but cannot provide proper film formation in boundary regimes. Solid lubricants can be applied as a component of anti-seize pastes and anti-friction coatings to offer protection under boundary conditions. These same solid lubricants can support lubrication solutions for the mixed regime. Effective lubrication for boundary regimes is needed in applications with static or heavy loads, components operating at low speeds, or with shock loads or transient conditions that might cause component failures. Anti-seize pastes and anti-friction coatings, fortified with high levels of lubricant solids, can go beyond traditional oils and greases to reduce wear; optimize friction; and provide long-lasting, effective lubrication in extreme environmental conditions.

The technology of solid lubrication has advanced rapidly in the past four decades, primarily in response to the needs of the aerospace and automobile industries. Solid lubricants are used where the containment of liquids is a problem and when liquid lubricants do not meet the advanced requirements. Under high vacuum (such as in space), high temperatures, cryogenic temperatures, radiation, dust, clean environments, or corrosive environments, and combinations thereof, solid lubrication may be the only feasible system. The materials designed for solid lubrication must not only display desirable coefficients of friction (0.001 to 0.3) but must maintain good durability in different environments, such as high vacuum, water, the atmosphere, cryogenic temperatures, high temperatures, or dust. Therefore, the successful use of materials and coatings as solid lubricants requires understanding their material and tribological properties and knowing which solid lubricant formulation is best for a chosen application. Issues such as substrate surface pretreatment, materials and coatings compatibility, the mating counterpart material, and potential debris generation must be taken into account during the design of a lubricated device or of moving mechanical assemblies.

Revision Questions

1. *What is a tribo-system?*
2. *With help of Stribeck curve explain hydrodynamic, boundary and mixed regimes of lubrication?*
3. *List main forms of lubricants and how they lubricate?*
4. *Name two forms of solid lubricants and explain how they lubricate under boundary conditions?*
5. *What are the applications of pastes and AFCs?*
6. *Give advantages as well as limitations of AFCs?*
7. *How does application of solid lubricants and AFCs revolutionize lubrication?*

References

Chichester C (2012) Lubrication beyond oil and grease, application engineering and technical service (AETS). Molykote, Dow Corning Corporation, USA

Miyoshi K (2007) Solid lubricants and coatings for extreme environments: state-of-the-art survey. National Aeronautics and Space Administration, Glenn Research Center, Cleveland, OH

Zhmud B, Pasalskiy B (2013) Technical note: nanomaterials in lubricants: an industrial perspective on current research, Lubricants 1(4):95–101, www.mdpi.com/journal/lubricants. doi:10.3390/lubricants1040095

Chapter 16
Conclusion

Abstract The chapter summarizes the key areas covered by the authors in this book, with contemporary issues being the topical use synthetic lubricants to replace mineral oils. The feasibility of eco-friendly bio lubricants for total loss applications in the view of thermal stability challenges, as well as the use of nano solid lubricants to replace fluids and grease for severe space operations. Reduction of emissions resulting from sulfur in additives is one way to save the environment among other efforts such as recycling of used oil and proper oil handling to minimize spills and leakages into the environment.

16.1 Future Lubricant Technology Drivers

The three pillars guiding engine lubricant technology advancement are the need to achieve significant fuel economy, appreciably extend oil drainage period and minimize gas emissions into the environment through sulfur-less additive packages. This is summarized by Fig. 16.1.

Consideration for use of vegetable based lubricants for total loss applications still faces high inhibiting costs for user compared to mineral based lubricants, and they do not seem to be readily acceptable to the market. Alongside this development has been the strides in synthetic hydrocarbon formulations to get biodegradable base oils (ATC 2007).

16.2 Synthetic Engine Oil Revolution

During its working life in an engine, synthetic lubricant oil accumulates various contaminants, including carcinogenic polycyclic aromatic hydrocarbons (PAHs), which are transferred from fuel as combustion products. Because oil change intervals are extended with synthetic lubricants, these used oils may contain higher

© Springer International Publishing Switzerland 2016
I. Madanhire and C. Mbohwa, *Mitigating Environmental Impact
of Petroleum Lubricants*, DOI 10.1007/978-3-319-31358-0_16

Fig. 16.1 Drivers of future oil formulations

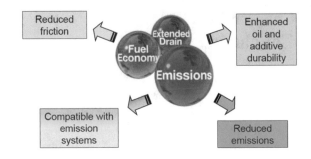

PAH concentrations than used conventional motor oils. Increased PAH levels in synthetic oil drained from vehicles and collected for recycling should be considered and controlled as appropriate when the used oil is burned or re-refined. Further, used synthetic oils that are leaked or spilled into storm water or other surface runoff may increase used oil-related PAHs in runoff and receiving waters.

Studies have reported higher PAH levels in particulate emissions from engines operating with used motor oil than with fresh lubricant oil. Because synthetic lubricants are likely to accumulate PAHs to a greater extent due to their extended life in the engine, it is likely that particulate emissions from engines using synthetic oils will have higher concentrations of PAHs. PAOs show higher biodegradability than mineral oils of equivalent viscosity because of their higher degree of hydrocarbon chain linearity. However, even within a class of synthetic lubricants, biodegradability ranges widely and various biodegradability tests can give different results for the same lubricant type (Bardasz 2005).

Performance characteristics of synthetic oils—friction reduction, engine parts wear minimization and lubricant chemical stability—act to decrease fuel consumption, help the engine to operate more efficiently and reduce greenhouse gas and other contaminant emissions. Recent studies suggest that synthetic lubricants yield lower levels of PM and other pollutants in engine exhaust emissions compared to mineral-based oils. As motor oil formulations transition to synthetics and other highly refined base oils, it is likely that the quality of the used oil pool available for recycling will improve. Using this as feedstock for re-refining will, in turn, likely yield a higher-quality, and more valuable re-refined base oil product.

16.3 Fate of Lubricant Additives

Crankcase lubricants are used to fill the vehicle sump at the recommended service interval, and for top-up in between. Lubricants leave the engine through lubricant consumption; down the tailpipe as gaseous emissions (combustion products or particulates), leakage through seals and gaskets during use, or in drained used oil at oil changes. Increasingly particulate traps are being fitted to vehicles which also capture ash derived from the lubricant. Below a number of final destinies for the oils noted in the prior chapters have been listed (Bardasz 2005).

Re-refining/re-use: Used crankcase lubricant is re-refined to get base stocks, which is then used, after appropriate treatment, in the same way as virgin base stocks. Re-refining and re-use prolongs the life of limited natural resources. Re-refining also produces a residual sludge. After appropriate treatment to neutralize it, the sludge is used in cement manufacture or road construction, or otherwise disposed of in landfills.

Incineration/use as fuel oil: Used oil is burnt either as fuel oil or used for heat recovery in incinerators. These activities provide significant environmental benefits, as it has been estimated that burning one ton of used oil saves the use of 0.85 ton of heavy fuel. These quantities include some 150 kilo tons of additives, of which no more than 7.5 kilo tons are elements other than carbon, hydrogen or oxygen, and not more than 4 kilo tons is metallic (calcium, magnesium and zinc). Incineration is regulated and monitored, and emissions originating from the additive content of used oils therefore present no new or unregulated environmental hazard.

Used oil/leakage: A lot of used oil is either unaccounted for or lost. At worst, this ends up in the water compartment, either directly or indirectly through the soil compartment. This used oil contains a lot of additives. The additives in used oil which could potentially enter the water compartment comprise mainly carbon, hydrogen and oxygen. The quantities involved do not represent a major environmental concern relative to total hydrocarbon leakages. However, dumping into the environment is unnecessary and unacceptable pollution which is being addressed by stricter implementation of existing regulations, and an increased focus on collection of used oil. Research has not yet identified alternative chemicals which offer the same benefits of reduced emissions, enhanced fuel economy and lower oil consumption, and could be considered more environmentally benign.

Fate of carbon: When fuel is burnt in cars and trucks, most of the carbon forms carbon dioxide. Small proportions end up in particulates and partially burnt hydrocarbon fractions. After-treatment devices will remove particulates and convert hydrocarbons and carbon monoxide to carbon dioxide (and water). Carbonaceous particulate matter will also be converted to carbon dioxide during the regeneration phase.

Other elements: It is assumed that all the additive metals burnt during combustion contribute to particulate formation. Magnesium and calcium form sulphates, whilst zinc goes to both zinc oxide and zinc pyrophosphate. The emissions of lubricant derived metals (S, Ca, Zn, P, and Mg) are highly correlated with emissions predicted from the composition of oil (and fuel sulphur); however, recovery rates vary considerably (ranging from 17 % for Mg to 125 % for S).

Phosphorus: The most common source of phosphorus is zinc di-alkyldithio-phosphate (ZDDP). In response to concerns about the impact of phosphorus on catalytic after-treatment systems, phosphorus levels in lubricants have been significantly reduced over the last decade.

16.4 Environmental Friendly Lubricants Reality

Recently, there has been a resurgence of eco-friendly lubricants due to increased environmental efforts to reduce the use of petroleum-based lubricants in addition to the depletion of oil reserves, increases in oil price, and rises in lubricant disposal costs. Although the lubrication market is shifting to become more environmentally responsible by reducing the use of petroleum-based lubricants due to concerns of protecting the environment, depletion of oil reserves, and increases in oil price, mineral oil remains to be the largest constituent and the foundation to most lubricants. When compared to petroleum-based lubricants, bio-based lubricants have a higher lubricity, lower volatility, higher shear stability, higher viscosity index, higher load carrying capacity, and superior detergency and dispersancy, therefore they are excellent alternatives to petroleum-based oils. Despite these favorable attributes, the largest drawbacks to many bio-based oils are their poor thermal-oxidative stability, high pour points, and inconsistent chemical composition, which have led to the development of chemically modified synthetic bio lubricants, the use of stabilizing additives, and ionic liquids (Reeves and Menezes 2016).

16.5 Concluding Remarks

The best solution is to opt for 100 % natural materials with such desired characteristics. In this way, great success can be attained in these industries. Further research on lubricant industries is suggested, specifically on the best method to reduce lubricant waste. Millions of tons of these materials are released into the environment annually. Reducing lubricant waste will not only save the environment and address inherent threats, but also facilitate economic growth, considering the cost of lubricant waste.

A majority of the natural additives and thickening agents used in lubricating grease formulations have contributed to the safety of the grease formulation process in terms of the natural biodegradable sources used and environmental considerations. These natural additives have been able to replace at least, if not fully extinguish, the toxicity associated with conventional additives. They are also very specific in their functions in greases, which include enhanced mechanical stability when polymers in the form of poly-olefins are added.

Although vegetable oils are a good alternative because they are environmentally friendly, their full replacement is not cost effective because of their low oxidative stability and other thermal and temperature considerations despite chemical modifications and mixing with other materials. Even after the efforts made to improve them, they remain incapable of meeting the standards. Such drawback makes them distinct from generally acceptable base oil, which meets all of the required criteria in terms of cost. Thus, more efforts must be exerted in this regard.

It is also time, leading companies in the lubricating sector should develop and scale-up innovative fully formulated lubricating oils incorporating nano-particles. Advanced nano materials, presently under study, have shown initial promising attitudes for reducing friction and enhancing protection against wear. Meaning there is a lot of potential to be tapped out of such research area.

Esters are assumed to be better base stocks than mineral oils in terms of performance, but they are more expensive and demonstrate poor wear resistance, sealant performance, and hydrolytic stability. By contrast, other vegetable oils can perform these functions when chemically modified, mixed with other base stocks such as natural triglycerides, or added with other components. Thus, if given attention, esters comprising the appropriate proportion of components can solve a number of inherent problems in the near future.

Lubricant technologists can deliver environmental benefit by extending lubricant lifetimes in order to reduce lubricant consumption, and by developing lubricant formulations with optimized friction and wear performance, while avoiding use of materials which may be persistent in the environment, bio accumulative or toxic.

Revision Questions

1. *What are the short comings of synthetics as future lubricants?*
2. *What is the serious handicap of vegetable oils?*
3. *What can be the contribution of grease on the environmental front?*
4. *What is the major impact of lubricants development on emissions?*
5. *What role can regulatory authorities have on lubricant usage and control?*

References

ATC—Technical Committee of Petroleum Additive Manufacturers in Europe (2007) Lubricant additives and the environment, ATC Document 49 (revision 1), Dec 2007

Bardasz EA (2005) Future engine fluids technologies: durable, fuel-efficient, and emissions-friendly, 11th Diesel Engine Emissions Reduction Conference, Chicago, 21–25 Aug 2005

Reeves CJ, Menezes PL (2016) Advancements in eco-friendly lubricants for tribological applications: past, present, and future. Eco-tribology research developments—materials forming, machining and tribology, vol 2. Springer International Publishing Switzerland, p 41–61. http://www.springer.com/978-3-319-24005-3, doi:10.1007/978-3-319-24007-7_2

Appendix

Refining Process of Mineral Base Oil

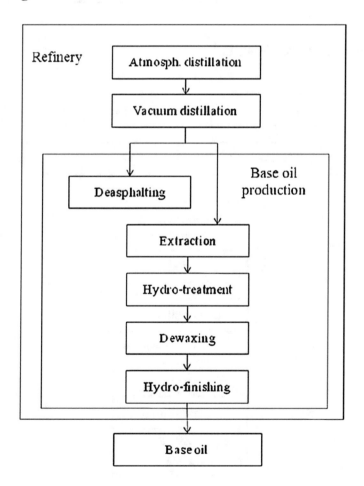

© Springer International Publishing Switzerland 2016
I. Madanhire and C. Mbohwa, *Mitigating Environmental Impact*
of Petroleum Lubricants, DOI 10.1007/978-3-319-31358-0

API Base Stocks Categorization Table

Group	Type	Saturates (wt. %)	Sulphur (wt. %)	VI	Manufacturing method	Relative cost
I	Mineral based	<90	>0.03	80–120	Solvent refining	1
II	Mineral based	≥90	≤0.03	80–120	Hydro-processing	2
III	Mineral based	≥90	≤0.03	>120	All hydro-processing	2.5
IV	Polyalpha-olefins (PAOs)	No	No	170–300	Chemical reaction	5
V	Synthetics	No	No		As indicated	2–4

Production of PAO Base Oil

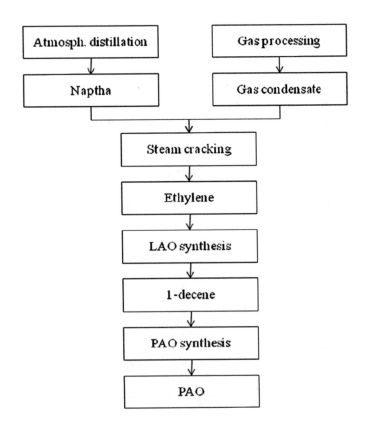

Typical Composition of Lubricating Oils Table

Compound	Weight percentage (%)
Base oil	86
Viscosity index improvers (polyisobutylene, polymethacrylate)	5
Oxidation inhibitor (zinc dialkyl, dithiophosphate)	1
Detergent (barium and calcium sulphonates or phenates)	4
Multi-functional additives (dispersant, pour point depressant)	4

Refining Options for Base Oil Production

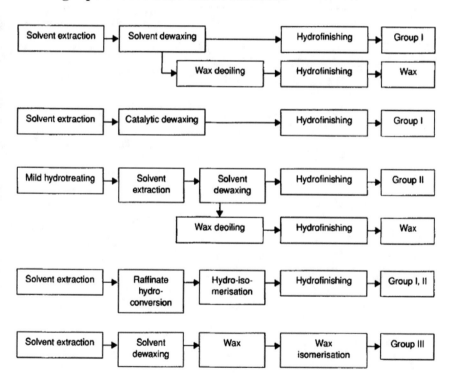

Schematic Lubricant Life Cycle

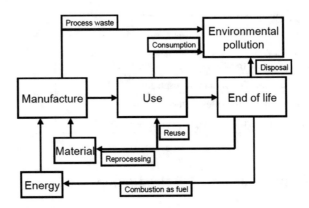

Glossary

Absorbent filter A filter medium primarily intended to hold soluble and insoluble contaminants on its surface by molecular adhesion

Acidification The problem of acidification is caused by acid depositions which originate from antropogenic emissions. The reference substance for the measurement of the acidification potential is SO_2. Acid depositions e.g., in the form of "acid rain" have a negative impact on water, forest, and soil

ASTM American Society for Testing Materialist is a society for developing standards for materials and test methods

Baffle A device to prevent direct fluid flow or impingement on a surface

Base oil A liquid product totally or partially consisting of mineral oil or synthetic fluid used as primary component for various type of marketed lubricants including engine oils automotive transmission fluids, hydraulic fluids, gear oils, metalworking oils, medicinal white oils and greases

Beta rating The method of comparing filter performance based on efficiency. This is done using the Multi-Pass Test which counts the number of particles of a given size before and after fluid passes through a filter

Bleeding The separation of some of the liquid phase from a grease

Blotter spot test A procedure used to determine whether a significant amount of sludge or varnish is present in the fluid

Breather A device which permits air to move in and out of a container or component to maintain atmospheric pressure

Bypass filtration A A system of filtration in which only a portion of the total flow of a circulating fluid system passes through a filter at any instant or in which a filter having its own circulating pump operates in parallel to the main flow

© Springer International Publishing Switzerland 2016
I. Madanhire and C. Mbohwa, *Mitigating Environmental Impact of Petroleum Lubricants*, DOI 10.1007/978-3-319-31358-0

Carcinogen A cancer-causing substance. Certain petroleum products are classified as potential carcinogens OSHA criteria. Suppliers are required to identify such products as potential carcinogens on package labels and Material Safety Data Sheets

Contaminant Any foreign or unwanted substance that can have a negative effect on system operation life or reliability

Cutting oil A lubricant used in machining operations for lubricating the tool in contact with the work piece and to remove heat. The fluid can be petroleum based, water based, or an emulsion of the two. The term "emulsifiable cutting oil" normally indicates a petroleum-based concentrate to which water is added to form an emulsion which is the actual cutting fluid

Filter Any device or porous substance used for cleaning and removing suspended matter from a gas or fluid

Filtration The physical or mechanical process of separating insoluble particulate matter from a fluid such as air or liquid, by passing the fluid through a filter medium that will not allow the particulates to pass through it

Fire resistant fluid Lubricant used especially in high-temperature or hazardous hydraulic applications. Three common types of fire-resistant fluids are: (1) water-petroleum oil emulsionsin which the water prevents burning of the petroleum constituent (2) water-glycol fluids and (3) non-aqueous fluids of low volatility such as phosphate esters, silicones, and halogenated hydrocarbon-type fluids

Flushing A fluid circulation process designed to remove contamination from the wetted surfaces of a fluid system

Foam An agglomeration of gas bubbles separated from each other by a thin liquid film. If an oil is said to not foamthe small air bubbles will quickly combine, become larger bubbles, and then break to vent to the atmosphere. If this action occurs slowly, the oil is said to foam

Gear oil A high-quality oil with good oxidation stability load-carrying capacity, rust protection, and resistance to foaming, for service in gear housings and enclosed chain drives. Specially formulated industrial EP gear oils are used where highly loaded gear sets or excessive sliding action (as in worm gears) is encountered

Hydraulic fluid Fluid serving as the power transmission medium in a hydraulic system. The most commonly used fluids are petroleum oils synthetic lubricants, oil-water emulsions, and water-glycol mixtures. The principal requirements of a premium hydraulic fluid are proper viscosity, high viscosity index, anti-wear protection (if needed), good oxidation stability, adequate pour point, good demulsibility, rust inhibition, resistance to foaming, and compatibility with seal materials. Anti-wear oils are frequently used in compact, high-pressure, and capacity pumps that require extra lubrication protection

Industrial lubricant Any petroleum or synthetic-base fluid or grease commonly used in lubricating industrial equipment such as gears, turbines, and compressors

ISO International Standards Organization sets viscosity reference scales

Lubricant Any substance interposed between two surfaces in relative motion for the purpose of reducing the friction and/or the wear between them

Lubricant A liquid product totally or partially consisting of mineral or synthetic oil that works to prevent metal-to-metal contact removes contaminants, cools machine surfaces, removes wear debris and transfers power. Lubricating oils are composed of base oils and additives

Lubrication The control of friction and wear by the introduction of a friction-reducing film between moving surfaces in contact. The lubricant used can be a fluid solid, or plastic substance

Materials safety data sheets A publication containing health and safety information on a hazardous product (including petroleum). The OSHA Hazard Communication Standard requires that an MSDS be provided by manufacturers to distributors or purchasers prior to or at the time of product shipment. An MSDS must include the chemical and common names of all ingredients that have been determined to be health hazards if they constitute 1 % or greater of the product's composition (0.1 % for carcinogens). An MSDS also included precautionary guidelines and emergency procedures

Mineral oil Oil derived from a mineral source as opposed to oil derived from plants or animals. The term is applied to a wide range of products that is typically used when referring to petroleum-based lubricants

Nutrification A process by which a body of water acquires a high concentration of plant nutrients especially nitrates or phosphates. Then nutrification promotes algae growth and can lead to a depletion of dissolved oxygen. Although nutrification is a natural process human activities can greatly accelerate nutrification

Oil Oil of any kind or in any form including, but not limited to, petroleum, fuel oil, sludge, oil refuse, and oil mixed with wastes other than dredged spoil

Oxidation Occurs when oxygen attacks petroleum fluids. The process is accelerated by heatlight, metal catalysts and the presence of water, acids, or solid contaminants. It leads to increased viscosity and deposit formation

PAO's (poly alpha olefins) synthetic base fluids for high performance lubricants. PAO's are impurity-free and contain only well-defined hydrocarbon molecules. They offer excellent performance over a wide range of lubricating properties. PAO's are manufactured by a two-step reaction sequence from linear α-olefins which are derived from ethylene

Patch test A method by which a specified volume of fluid is filtered through a membrane filter of known pore structure. All particulate matter in excess of an "average size" determined by the membrane characteristics, is retained on its surface. Thus, the membrane is discolored by an amount proportional to the particulate level of the fluid sample. Visually comparing the test filter with standard patches of known contamination levels determines acceptability for a given fluid

Pour point Lowest temperature at which an oil or distillate fuel is observed to flow when cooled under conditions prescribed by test method ASTM D 97. The pour point is 3 °C (5 °F) above the temperature at which the oil in a test vessel shows no movement when the container is held horizontally for five seconds

Regeneration Any process whereby base oils can be produced by refining waste oils in particular by removing the contaminants, oxidation products and additives contained in such oils

Release Any spilling leaking, emitting, discharging, escaping, leaching, or disposing from an UST into groundwater, surface water, or subsurface soils

Sludge Insoluble material formed as a result either of deterioration reactions in an oil or of contamination of an oilor both

Stoke (St) Kinematic measurement of a fluid's resistance to flow defined by the ratio of the fluid's dynamic viscosity to its density

Synthetic lubricant A lubricant produced by chemical synthesis rather than by extraction or refinement of petroleum to produce a compound with planned and predictable properties

Turbine oil A top-quality rust- and oxidation-inhibited (R&O) oil that meets the rigid requirements traditionally imposed on steam-turbine lubrication. Quality turbine oils are also distinguished by good demulsibility a requisite of effective oil-water separation. Turbine oils are widely used in other exacting applications for which long service life and dependable lubrication are mandatory. Such compressors, hydraulic systems, gear drives, and other equipment. Turbine oils can also be used as heat transfer fluids in open systems, where oxidation stability is of primary importance

Underground storage tank Any one or combination of tanks (including underground pipes connected thereto) that is used to contain an accumulation of regulated substances and the volume of which (including the volume of underground connected thereto) is 10 % or more beneath the surface of the ground

Varnish When applied to lubrication a thin, insoluble, non-wipeable film deposit occurring on interior parts, resulting from the oxidation and polymerization of fuels and lubricants. Can cause sticking and malfunction of close-clearance moving parts. It is similar to, but softer, than lacquer

Viscosity Measurement of a fluid's resistance to flow. The common metric unit of absolute viscosity is the poise which is defined as the force in dynes required to move a surface one square centimeter in area past a parallel surface at a speed of one centimeter per second, with the surfaces separated by a fluid film one centimeter thick. In addition to kinematic viscosity, there are other methods for determining viscosity, including Saybolt Universal Viscosity (SUV), Saybolt Furol viscosity, Engler viscosity, and Redwood viscosity. Since viscosity varies in inversely with temperature, its value is meaningless until the temperature at which it is determined is reported

Waste oil Any mineral-based lubrication or industrial oil which have become unfit for the use for which they were originally intended

Printed in the United States
By Bookmasters